Eden Turned on Its Side

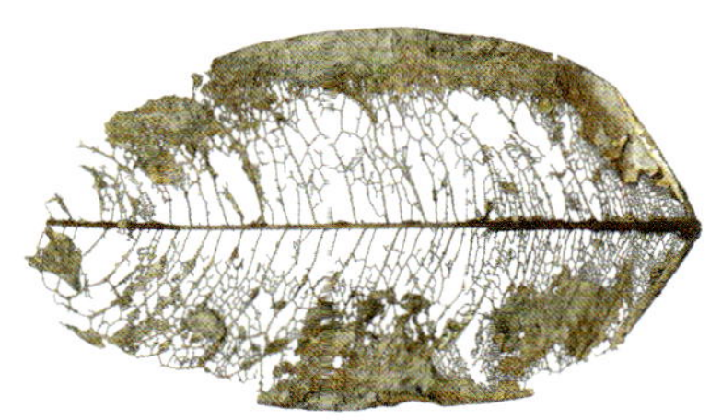

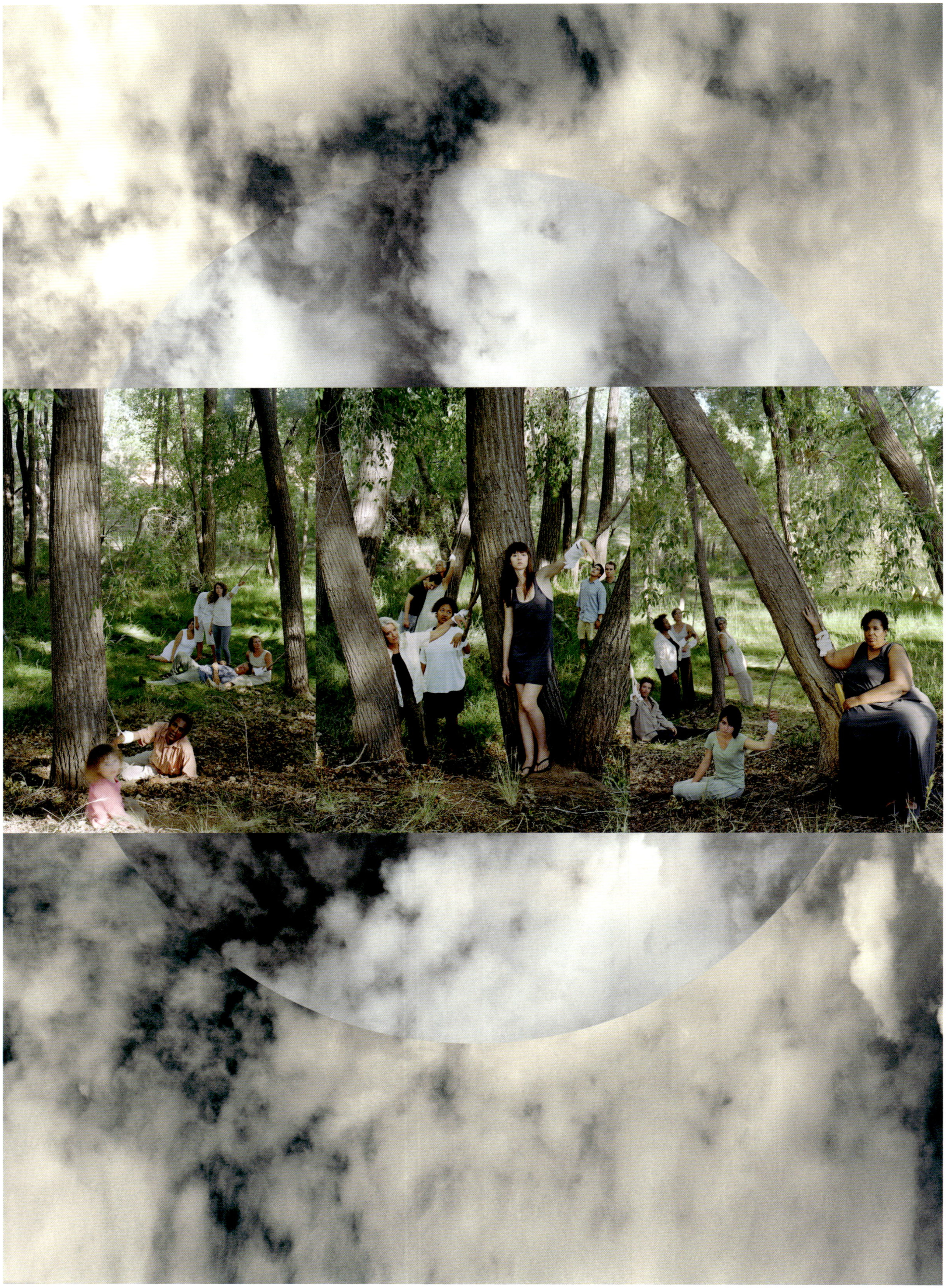

Eden Turned on Its Side

PHOTOWORKS BY Meridel Rubenstein

FOREWORD BY Arif Khan

ESSAYS BY Shawn Michelle Smith and Alan Weisman

UNIVERSITY OF NEW MEXICO ART MUSEUM • ALBUQUERQUE

Summer Solstice, 2009–2011, 54 ¼ × 40 in.

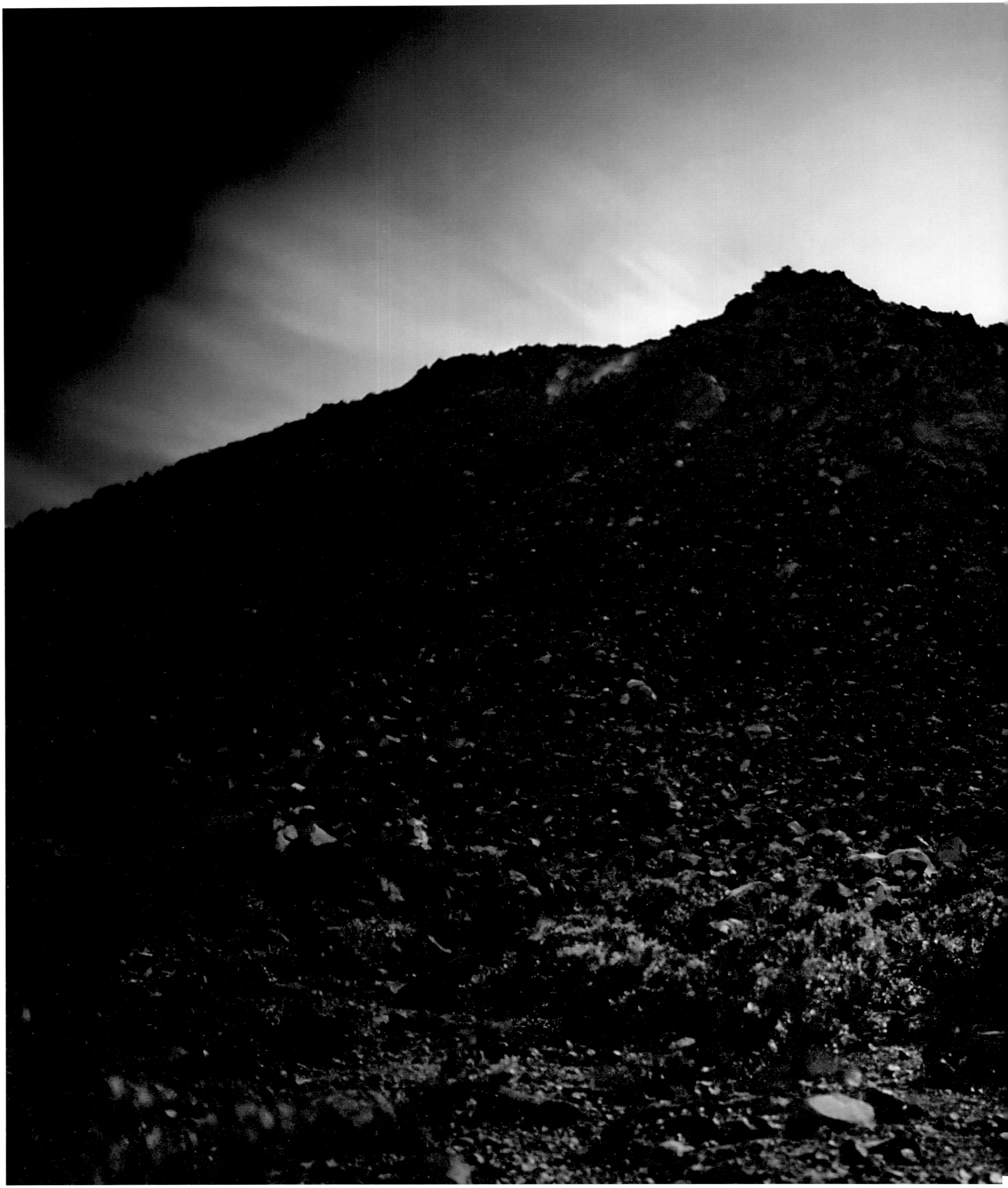

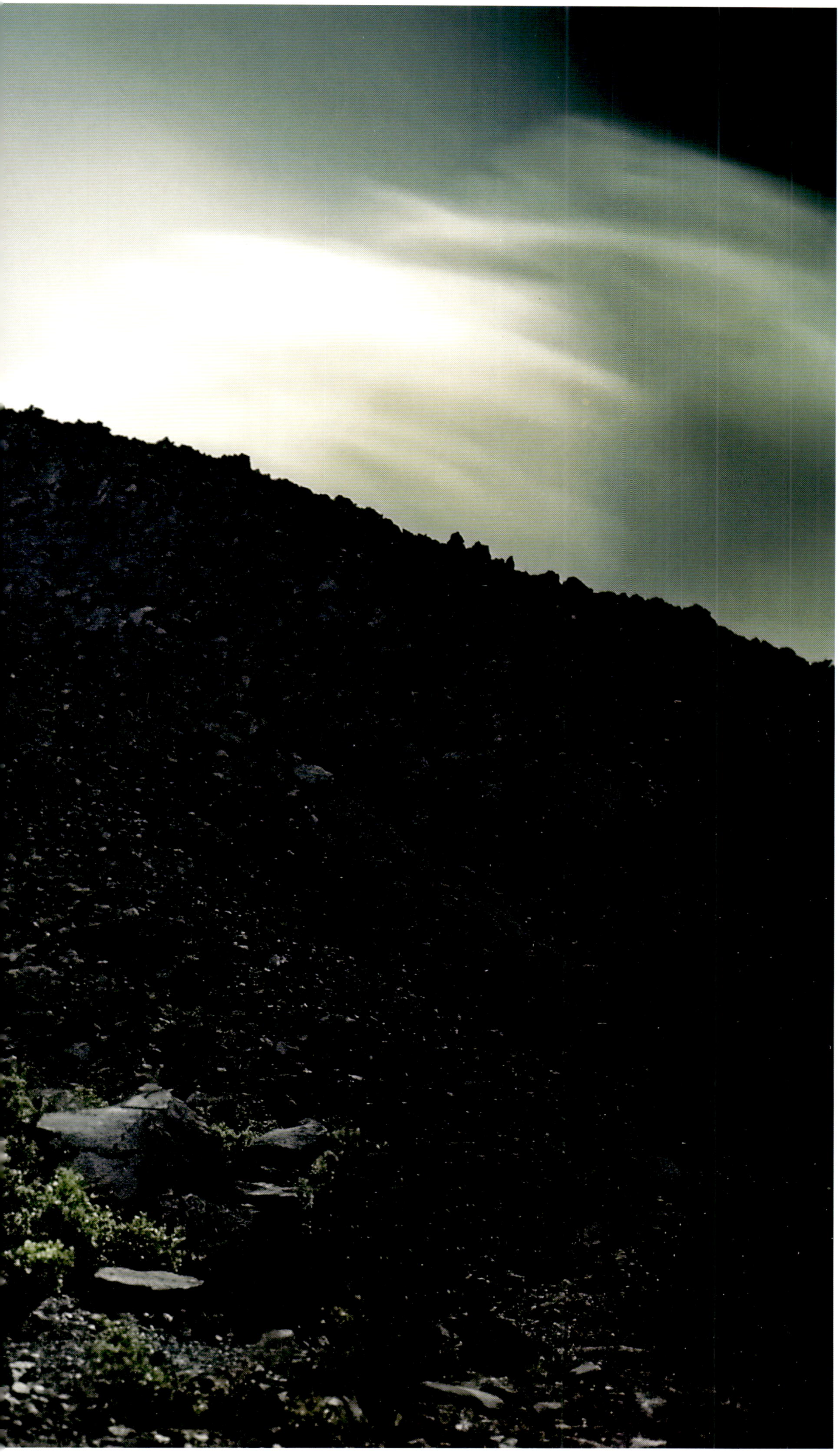

Mt. Merapi in Moonlight, The Calm before the Storm, Java, Indonesia, 2010, 18 ⅝ × 27 in.

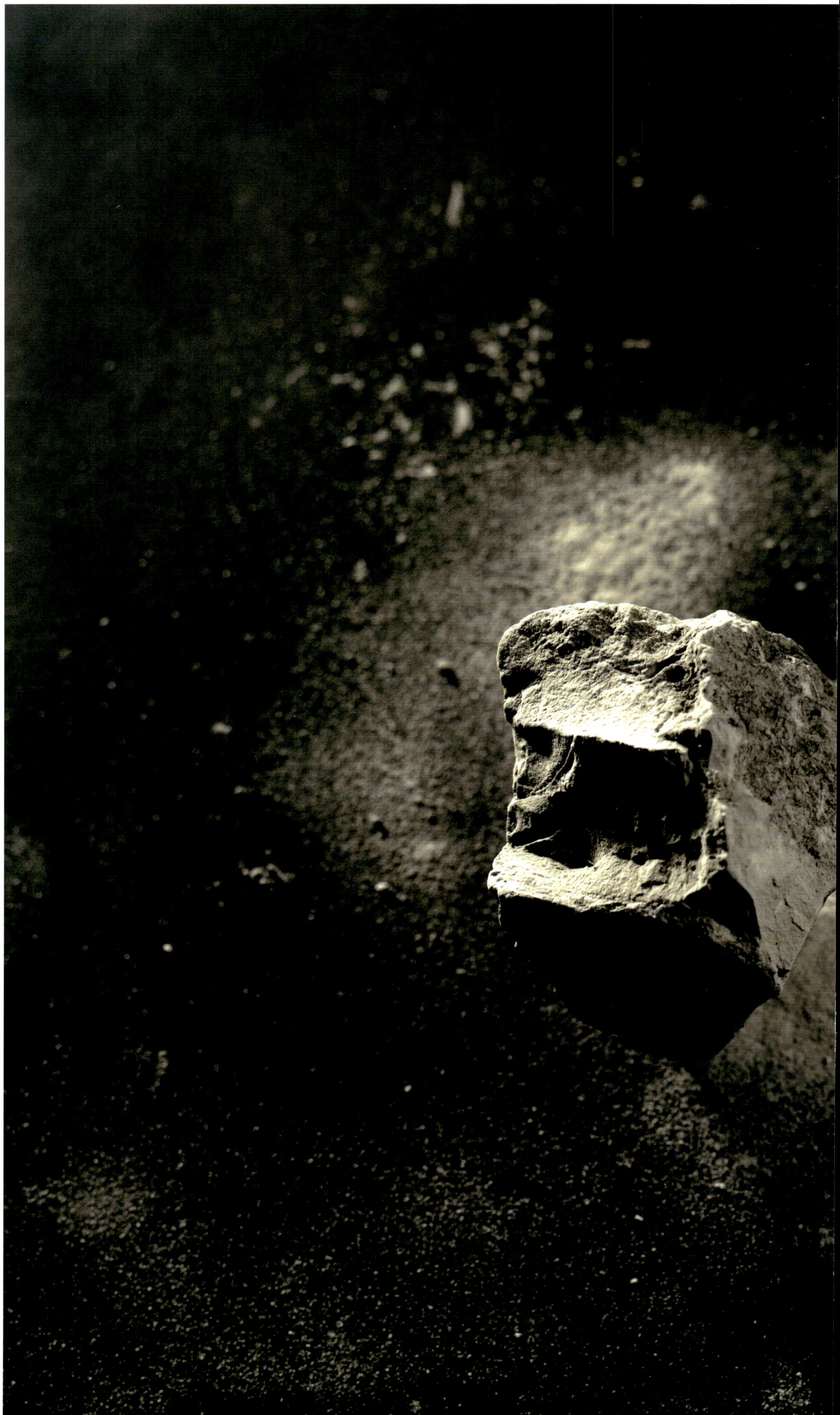

Mt. *Toba Volcanic Ash, 74,000 Years Old*, from Sumatra, found in Malaysia, 2010, 18 ⅝ × 27 in.

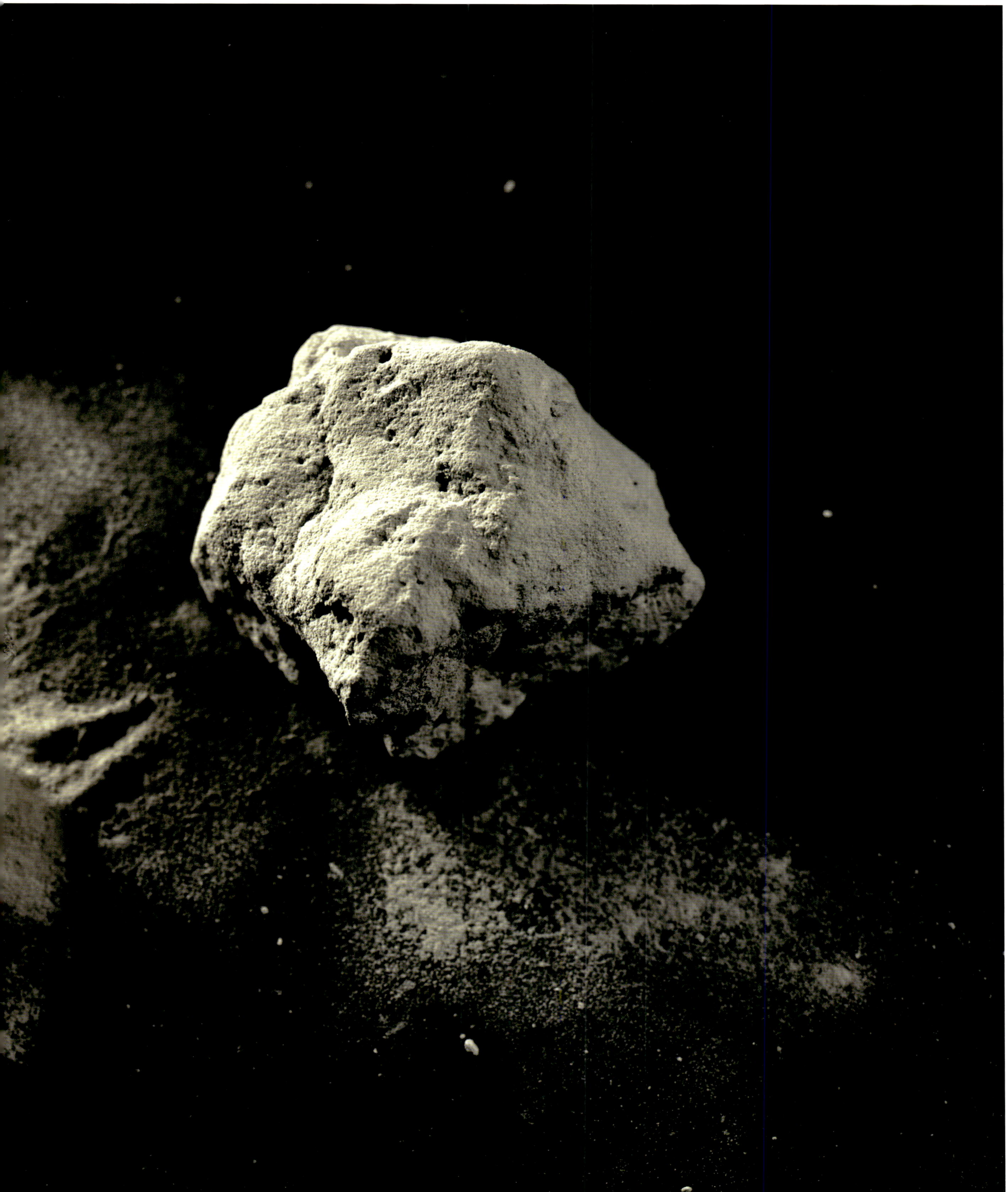

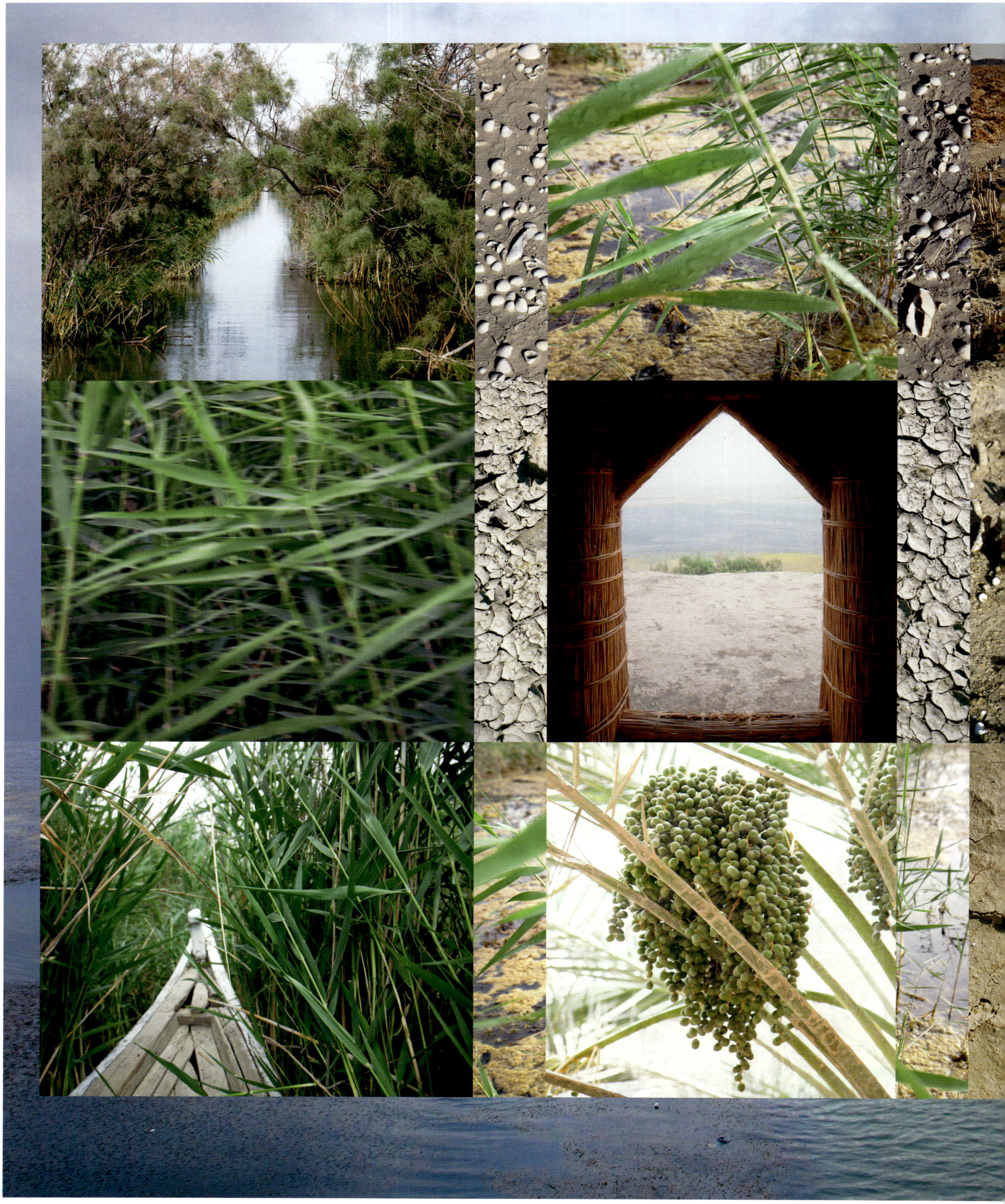

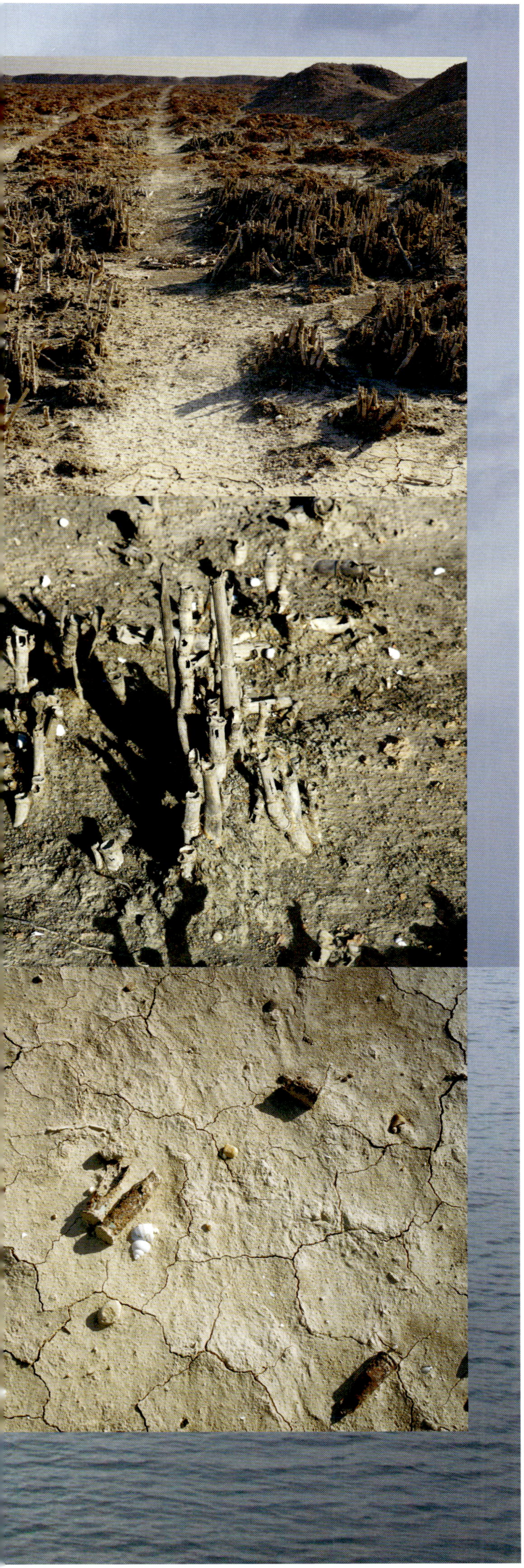

Contents

Eden Again, 89 × 74 in.

Foreword

Arif Khan

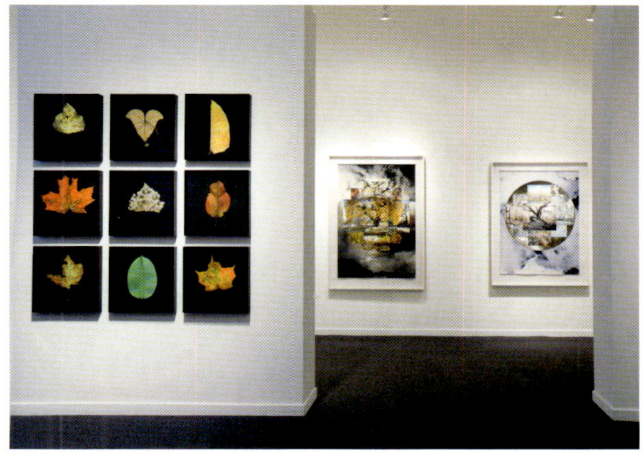

Eden Turned on its Side: Photosynthesis, Brian Gross Fine Art, San Francisco, California, 2012, installation view

As a teaching museum visited by students, faculty, scholars, and the general public, the University of New Mexico Art Museum in Albuquerque invites these diverse audiences to engage with a remarkable permanent collection of over thirty thousand works of art. The museum's designation as a teaching museum signals its intent to make collections, exhibitions, and museum use by students and faculty a significant aspect of the education it offers. In this educational setting, featuring the artwork of accomplished alumni is especially gratifying. Photography alumna Meridel Rubenstein, who came to study at the University of New Mexico in 1973, has since created an accomplished career as an artist and arts educator, and it is our honor to present this publication and exhibition, *Eden Turned on Its Side*.

The museum's collection of over ten thousand photographs began in the early 1960s under the leadership of Clinton Adams, the dean of the College of Fine Arts, and founding director Van Deren Coke. In less than a decade, the University of New Mexico established itself as a burgeoning center for the study of photography as studio practice and as a subject for art historians. The 1971 appointment of art historian Beaumont Newhall, former director of the George Eastman House in Rochester, New York, and first curator of photographs at the Museum of Modern Art, signaled the national reputation that the university had swiftly acquired. This exciting environment drew photographer Meridel Rubenstein from the east coast, and in four years she earned both an MA and an MFA in studio art and in the history of photography, respectively.

Eden Turned on Its Side presents the places where human culture and our natural environment connect and poetically displays the cultural value of nature. Rubenstein's focus is not so much on the landscapes that humans have left their mark on, but rather the landscapes that have left their mark on humans. This natural landscape has been the vast source from which our cultural inspirations and innovations have evolved. The modern pictorial history of Eden was produced by some of the most renowned artists in the history of art. However, the lineage stalls at the threshold of abstract expressionism and, seemingly, excludes photographic depictions. Such a finding raises a variety of questions: Is Eden still pictured today? Can it be created? Can it be photographed?

For Meridel Rubenstein the answer is an emphatic yes. Yes, Eden is still pictured today and can be found in the world at hand. Yes, Eden can be created, and it can be photographed. Rubenstein allows us entry to an infinite number of "Edens" if only we look attentively.

I would like to thank the three writers, guest curator Shawn Michelle Smith, author and environmental journalist Alan Weisman, and Meridel Rubenstein for their thoughtful contributions to this publication.

Final and lasting thanks to Meridel Rubenstein whose artworks and career have inspired this publication.

Arif Khan
Director
University of New Mexico Art Museum

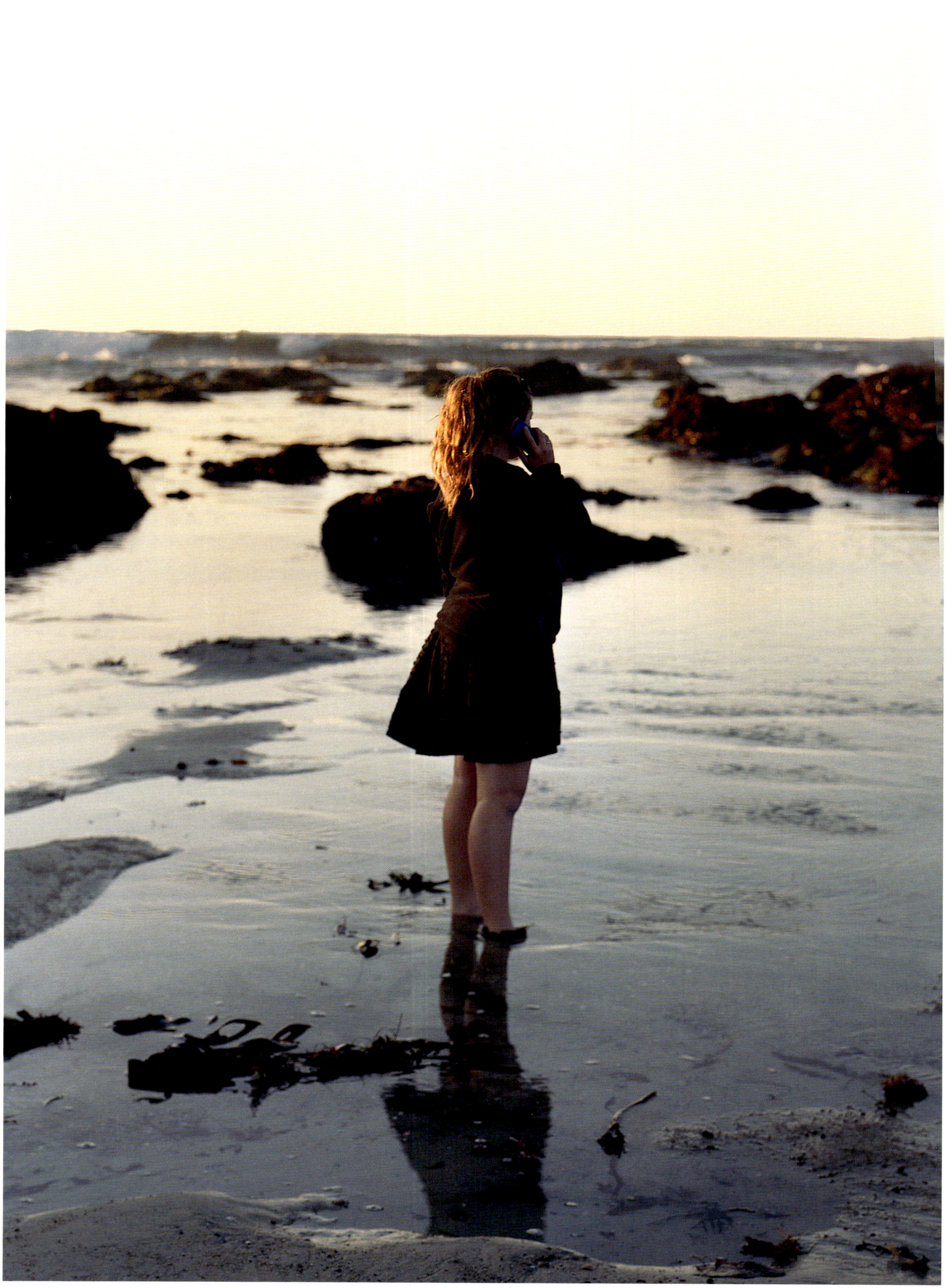

Tending the Garden

Shawn Michelle Smith

Meridel Rubenstein's *Eden Turned on Its Side* is about human relationships to the environment. It develops a theme that has long inspired Rubenstein, namely our intimate connection to the land, but it extends the line of inquiry temporally back into geological time and spatially out into a global expanse. The work consists of large-scale photographic images that tend toward installation pieces, as is characteristic of much of Rubenstein's art. It comprises three parts, *Photosynthesis*, *Volcano Cycle*, and *Eden in Iraq*, which explore ecologies on the scales of human time, geological time, and mythical time. The work resonates with current conversations about the Anthropocene, a newly named geological period in which the earth registers indelibly the mark of human industry, and especially the effects of fossil fuels. This is literally an "age of man" in which human life on the planet has inexorably transformed it, leading to the current crisis of climate change. Rubenstein's *Eden Turned on Its Side* engages this enormous transformation but asks viewers to consider their own place in this ongoing dynamic and their own relationship to the natural world. Entering into conversations that tend to universalize as they take a global view, Rubenstein's work invites one to look at the problem up close as well as far away and to consider what one might do, in even the smallest way, to repair the damage done. *Eden Turned on Its Side* locates viewers in the Anthropocene as agents rooted in particular landscapes that are connected on a global scale. It asks one to consider one's relationship to place and plants and people, both intimately and planetarily.

Photosynthesis

Photographers, like plants, turn to the sun—and photography, like photosynthesis, is a chemical process initiated by sunlight. Rubenstein's *Photosynthesis* focuses on the cycle of the seasons, the patterns of transformation in sky and earth that occur as spring grows into summer, summer shades into fall, fall collapses into winter, and winter thaws into spring. Some of the photographs focus on dazzling cloudscapes, while others piece together seasonal views around stately trees, and still others show people engaged in rituals of restoration and repair.

The earth that has sustained human life began with photosynthesis. Three billion years ago blue algae transformed the planet. According to Christophe Bonneuil and Jean-Baptiste Fressoz, "As the first living creatures to practice photosynthesis, they fixed carbon from the atmosphere into sediment in the ocean depths and released oxygen into the air, making it possible for the animals that appeared later to breathe, and forming the ozone layer that protects the planet from highly mutagenic ultraviolet radiation."[1] The very possibility of human life on earth began eons ago with photosynthesis, and this chemical process continues to generate the air we breathe.

The Anthropocene might also be said to have its roots in photosynthesis. Once again the term names a new geological period in which the earth has been permanently transformed by human production. Although attempts to date the advent of this new epoch remain contested, many historians propose the invention of the steam engine in 1784 and the subsequent industrial revolution of the nineteenth century as important beginnings. The fossil fuels that fed the industrial revolution, namely the coal, oil, and gas used for heating and manufacturing, were derived from carbon produced over millions of years of photosynthesis.[2] The chemical process that has sustained life on earth

Conversation with the Sea, 2004, 27 × 18 in.

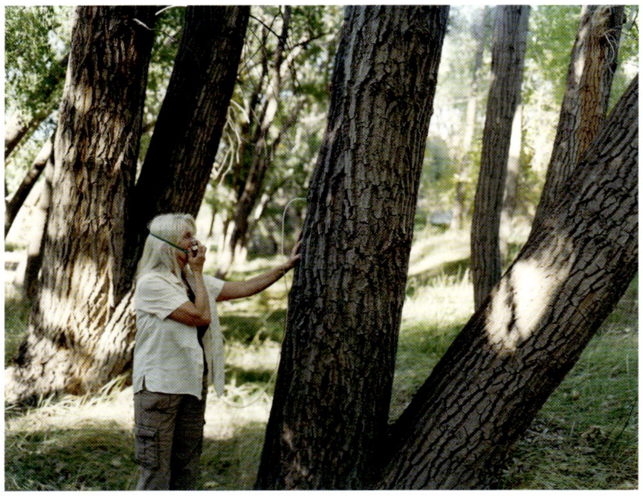

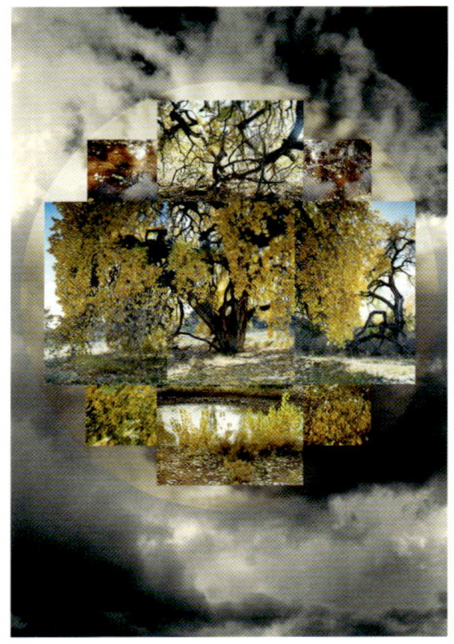

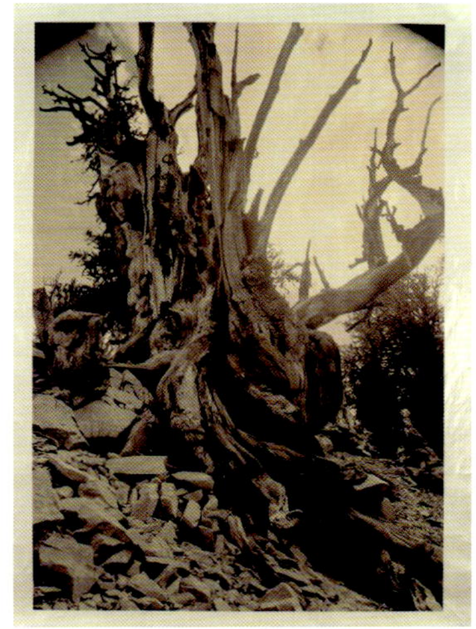

also created the stores of energy humans have used to endanger that life as we know it.

Rubenstein's *Photosynthesis* explores our physical dependence on plants to transform carbon dioxide into oxygen, and a few of the large-scale photographs envision this relationship directly. In *Respiration* a white-haired woman stands in the shade of a grove of trees. With one hand she holds an oxygen mask to her face, and with the other she reaches out to touch the trunk of a cottonwood tree. Looking closely one sees that the tubing of her oxygen mask is attached to the tree, which provides her life support. The photograph is disquieting in its directness. But the discomfort it generates is also instructive because the image asks one to consider questions that are uncomfortable, questions about our dependence on and integration with the natural world. *Respiration* underscores that as the earth ails, we do too.

In a number of Rubenstein's *Photosynthesis* images, trees function as pivots around which ecosystems revolve and coalesce. Deciduous trees mark the seasons as leaves bud, turn from green to yellow, and eventually fall from branches. In *Fall Seasonal*, nine rectangular images cohere around the trunk of a tree arrayed with golden leaves. Small images on the top and bottom present a variety of plants in fall foliage. The images are held together in a circular frame that recalls the revolving cycles of the seasons and the shape of the earth. The orb is situated within a larger rectangular frame of dramatic clouds that gestures to the atmosphere of the planet and the cosmos beyond.

Trees stand as witnesses to human and natural history in *Photosynthesis* as they have in other works by Rubenstein. *Millennial Forest*, for example, puts human life into perspective relative to living things that persist for thousands of years. In this work Rubenstein photographed the oldest trees in Vietnam and the United States, finding sentinels that both predated and outlived the war in Vietnam. As Rubenstein has said of the work, "The millennial trees in these photographs were not under siege when the images were made. Time has passed; their roots have continued to grow. They are survivors, and reminders that we share a common ground."[3] *Cousin of Methuselah*, for example, documents a Bristlecone Pine that grows near Bishop, California, and is roughly 4,700 years old. Rubenstein made the prints in this series on tree-bark paper coated with ground mica dissolved in gum arabic (resin), forging material connections between the photographs and their subjects. Photographs, like the ancient trees Rubenstein has studied, also endure beyond individual human lives, and Roland Barthes, one of photography's most influential commentators, found that fact startling. Indeed, in the persistence of the photograph across

from left to right:

Respiration, 2009–2011

Fall Seasonal, 2009–2011

Cousin of Methuselah, 2000

time Barthes saw only death: "The photograph tells me death in the future. . . . Whether or not the subject is already dead, every photograph is this catastrophe."[4] For Rubenstein the continued existence of the photograph beyond the life of its human subject is not so disturbing. Indeed, for her, photographs, and more importantly trees, which have lent their pulp to countless photographic prints, refer to life, not death, and to life that persists beyond the span of human hubris.

Volcano Cycle

If *Photosynthesis* is about the cyclical time of the seasons, *Volcano Cycle* is about the deep time of geology and the violent transformation of the world. *Volcano Cycle* documents the Indonesian Ring of Fire to consider environmental change on the nonhuman scale of geological time. In this series Rubenstein has photographed active volcanoes in Java, including Mount Bromo and Mount Merapi. She has produced the images on aluminum panels, giving the pieces a material heft and metallic sheen that resonates with the pyroclastic substances they record.

In *Lelani at Dawn, Mt. Bromo* a small figure walks across an ashen landscape. She is overshadowed by a giant plume that erupts from a mountain behind her. The image is both beautiful and unnerving. The erupting volcano might be understood to signal the awesome and destructive power of Gaia, the Greek goddess of the earth. The mountain range is bathed in light, but the foreground where Lelani walks is cast in shadow. The image suggests that a new era on earth is dawning, and one that will have unknown effects for its human inhabitants.

Geological time was conceived in the nineteenth century. In the 1830s geologists proposed a radical new timeline for the earth stretching back billions of years,

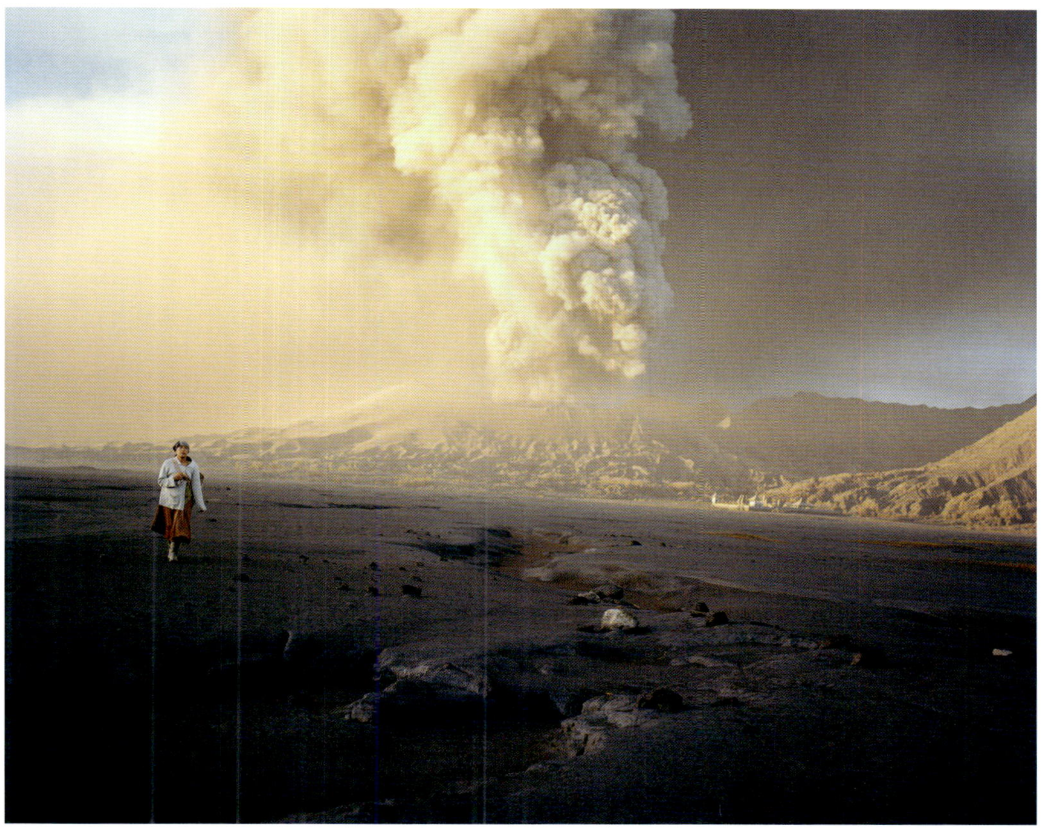

Lelani at Dawn, Mt. Bromo, 2011

and they thereby set the earth on a course that diverged from human and Biblical history. According to Bonneuil and Fressoz, "Whereas the Biblical account, like many other non-Western origin myths, made it possible for a long time to view human history as closely linked to that of the Earth . . . these two domains became increasingly separated in the course of the nineteenth century, as the prehuman history of the Earth became longer."[5] As geologists increasingly distinguished the history of the earth from that of humankind, political theorists and philosophers of the late eighteenth and nineteenth centuries also conceived of Nature as separate from man, as a realm to be dominated and a resource to be utilized.

The invention of photography was coincident with new conceptions of the history of the earth, as Robin Kelsey has argued: "Photography arrived roughly alongside the notion of geologic time. . . . The publication of Charles Lyell's *Principles of Geology* in three volumes between 1830 and 1833 did much to supplant the biblical account of creation with a story of geologic gradualism. A few years later, in early 1839, experimenters in France and England announced the invention of photography."[6]

With *Volcano Cycle*, Rubenstein binds photography with the deep time of geology. *Mt. Toba Volcanic Ash* presents two mounds covered with light, powdery dust that spills onto an indeterminate dark background. Although scale is difficult to judge, the image is almost forensic in its straightforward documentation, even as it is also aesthetic in its minimal forms, dramatic lighting, and high contrast. The image points to the "Toba catastrophe theory" according to which "a massive volcanic eruption changed the course of human history by severely reducing the human population." First presented by Stanley H. Ambrose in 1998, the theory proposes that a "mega-colossal" volcanic eruption in the Toba caldera in Indonesia approximately seventy-four thousand years ago produced massive environmental change. Pyroclastic material and gases shot into the atmosphere created a global volcanic winter that led to planetary cooling and eventually to the extinction of all other human species except for the branch that became modern humans.[7] Although it has since been challenged, the Toba catastrophe theory stands as a dramatic figuration of mass extinction and human evolution. Rubenstein's *Mt. Toba Volcanic Ash* encourages

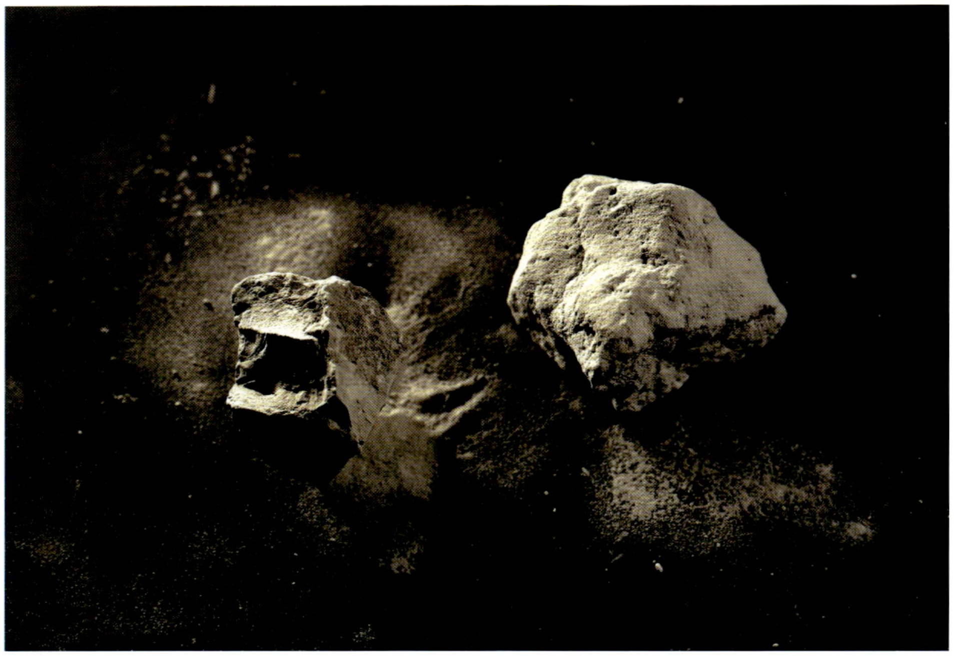

Mt. Toba Volcanic Ash, 74,000 Years Old, 2010

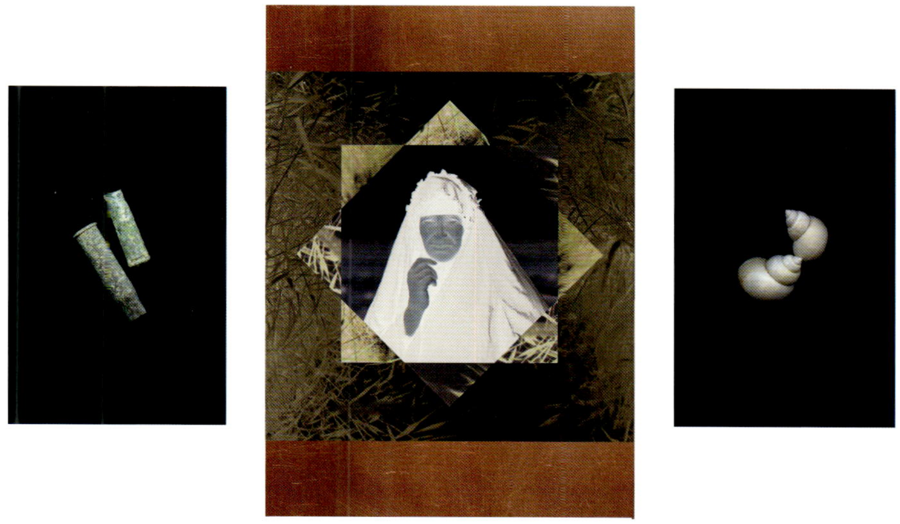

Study for *Temple of Inanna*, 2011–2017

one to consider both our potential evolutionary link and our ultimate vulnerability to cataclysmic geological events. The powdery ash of Mt. Toba figures both our beginning and our end.

Rubenstein grounds her volcanic imagery in the material of the earth by printing the images in *Volcano Cycle* on aluminum, the most abundant metal in the earth's crust.[8] She also encourages one to think about the material makeup of photography—its basis in chemistry and geology—even in the digital age. The aluminum plate of *Mt. Toba Volcanic Ash* might recall the copper plate of the daguerreotype, one of the first photographic technologies, or perhaps the less expensive plate of the slightly later tintype. Pursuing for photography what Jussi Parikka has called "a media history of matter,"[9] one might further contemplate the silver nitrate of early developing processes, the silver embedded in twentieth-century gelatin-coated papers, and the metallic and chemical composition of screens in the digital age.[10] In other words, one might consider photography's own geology.

Today, with the naming of the Anthropocene, human history and geological time are being reunited.[11] Nature can no longer be considered separate from human endeavor. Indeed, the nineteenth century, the very period in which geologists began to separate the two timelines, has been identified as a possible origin point for their conjoined trajectories. The escalating use of fossil fuels during the industrial revolution produced accelerating concentrations of atmospheric carbon dioxide, one of the chemical signatures scientists measure to mark the advent of the Anthropocene. Photography entered into these diverging and converging histories in the nineteenth century with its own geological formations.

Eden in Iraq

The Garden of Eden is the place from which Rubenstein renarrates the conjoined histories of earth and humanity. She takes us back to a Biblical conception of the earth created coterminously with humankind. Returning to a religious understanding dominant before the conception of geological time in the nineteenth century, she asks viewers to imagine the convergence of mythical, historical, and geological time. But in her work, Eden has been turned on its side. This is not the Garden conceived by a divine being, but the Anthropocene, the world that man has made.

The series *Eden in Iraq* explores environmental devastation and renewal in the marshes of southern Iraq, where the Tigris and Euphrates rivers meet, near the site of Biblical Eden on earth. In 1991, after popular uprisings that emerged during a cease-fire of the Gulf War, Saddam Hussein drained the Iraq wetlands in retaliation,

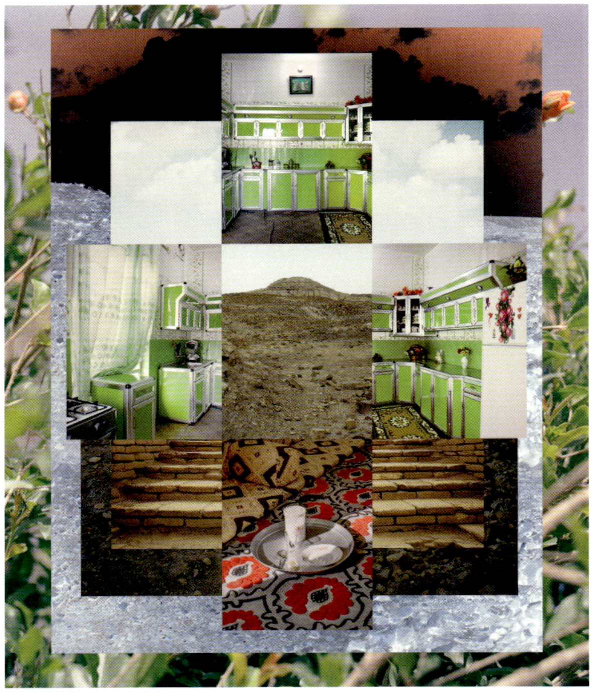

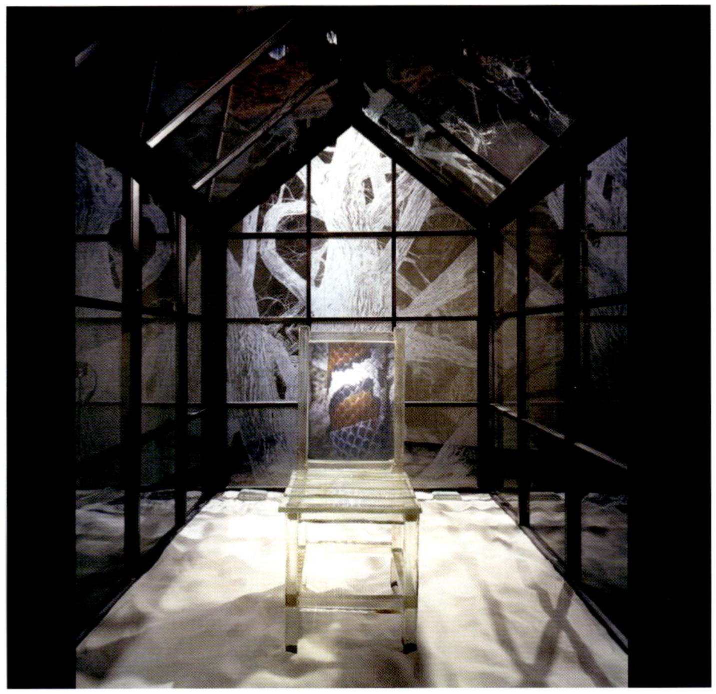

turning it into a desert and displacing the Marsh Arabs. Twelve years later, the US war in Iraq, which many critics deemed a "war for oil," wrought military destruction across the land and reinvigorated the demand for fossil fuels in the United States. Rubenstein's *Temple of Inanna* evokes this double history of environmental devastation in Iraq, mirroring bullet shells with the shells of marsh snails. These two different kinds of casings flank an image of a woman who represents the goddess Inanna, set in the grasses of a renewed marshland. Inanna, goddess of love and war, was patron deity of the city of Uruk and the most revered goddess of ancient Mesopotamia.[12] Responsible for "the redistribution of resources and fertility of the earth,"[13] Inanna serves as an appropriate symbol for the regreening of the Iraq marshlands in *Eden in Iraq*. In *Temple of Inanna* Rubenstein presents Inanna in negative, as if she is a latent power waiting to reemerge and restore the marshes.

War has historically been one of the destructive vehicles through which the world is conceived globally, and US involvement in wars of aggression has been a central theme in Rubenstein's work. In addition to the US war in Iraq, she has produced art about Vietnam as well as the Cold War. Indeed, the Cold War figures prominently in Rubenstein's *Oppenheimer's Chair* (1995), an installation and video piece that refers to J. Robert Oppenheimer, the physicist who was a key figure in designing the atomic bomb at Los Alamos National Laboratories in New Mexico, not far from where Rubenstein lives. The atomic bomb has left its own indelible mark on earth systems, and as Joseph Masco has argued, nuclear fallout functioned as an important register through which the earth was conceived globally during the Cold War. "By tracking the distribution of radioactive elements produced by atomic testing through the atmosphere and across plant, animal, and human populations, Cold War scientists were able to document the integration of the global biosphere."[14]

As Rubenstein considers the ways in which military offensives and scientific efforts have linked people globally and effected the environment planetarily, she also tacks back to the local scale, grounding some of her investigations in the lived space of the home. In *The Green Kitchen* the lime-green cabinets of a kitchen are

from left to right:

The Green Kitchen, 2014–2017

Oppenheimer's Chair, 1995

framed by images of dry, baked earth. These images are framed again by images of the regreening marshes. The bright green of the kitchen cabinets recalls the renewing landscape of the marshes outside, and the piece subtly links daily sustenance to ecological regeneration. Produced on linen to suggest a tapestry, the work implies that daily human needs are literally intertwined with larger earth systems.

Rubenstein's series *Eden in Iraq* corresponds to a larger project in which she is building a literal garden in the town of El Chibaish, across from the new Mesopotamian Marshes National Park in southern Iraq, recently designated a World Heritage Site. Sponsored by the NGO Nature Iraq, this environmental artwork is part of a major regreening effort underway in the region. Here Rubenstein will construct a wastewater garden to process sewage from the town while providing water for the garden. As a literal regeneration of the marshes and a symbolic regeneration of the Garden of Eden, the site will be an emblem of repair in the age of the Anthropocene.

Conclusion

Rubenstein's work asks one to imagine a new coeval relationship to the natural world that humans have utterly transformed. *Eden Turned on Its Side* underscores climate crisis and foreshadows the possibility of massive extinction, but it also holds out the possibility of renewal. The work grounds people in specific places within planetary environments and encourages one to recognize the ways in which daily life is imbricated in global forces. In doing so it also suggests that regeneration on the local level might effect transformation of the vast earth systems that one also inhabits. Although it may be impossible to right an overturned Eden in the age of the Anthropocene, at the very least one can begin to tend the garden at home.

Notes

1. Christophe Bonneuil and Jean-Baptiste Fressoz, *The Shock of the Anthropocene: The Earth, History and Us* (2015), trans. David Fernbach (London: Verso, 2016), 58.

2. Jussi Parikka, *A Geology of Media* (Minneapolis: University of Minnesota Press, 2015), 18, 50.

3. Meridel Rubenstein, *Belonging: Los Alamos to Vietnam* (Los Angeles: St. Ann's Press, 2004), 185.

4. Roland Barthes, *Camera Lucida: Reflections on Photography* (1980), trans. Richard Howard (New York: Hill and Wang, 1981), 96.

5. Bonneuil and Fressoz, *Shock of the Anthropocene*, 27.

6. Robin Kelsey, *Photography and the Art of Chance* (Cambridge: Belknap of Harvard University Press, 2015), 7. According to Kelsey, by proposing that the earth had come into being without Divine design, geology threatened to displace God as the maker of the world. In a parallel sense, photography, for nineteenth-century commentators, threatened to displace the artist as the maker of pictures, because the camera seemed to allow nature to represent itself through sunlight. In the convergence of geology and photography in the nineteenth century, then, one might see in a photographic landscape an image that appeared to make itself of a world that also seemed to make itself.

7. https://www.sciencedaily.com/terms/toba_catastrophe_theory.htm (accessed November 10, 2016) The theory was first presented by anthropologist Stanley H. Ambrose in his article "Late Pleistocene human population bottlenecks, volcanic winter, and differentiation of modern humans," *Journal of Human Evolution* 34 (1998): 623–51. The theory has since been challenged. https://anthropology.net/2007/07/06/mount-toba-eruption-ancient-humans-unscathed-study-claims/ (accessed November 10, 2016)

8. "Aluminium is the third most abundant element in the Earth's crust (after oxygen and silicon) and its most abundant metal." https://en.wikipedia.org/wiki/Aluminium (accessed November 10, 2016)

9. Jussi Parikka, *Geology of Media*, 25.

10. According to Parikka, W. Jerome Harrison's *History of Photography* (1887) might be read as "a story of chemicals instead of merely the inventor-experimenters such as Niepce, Daguerre, or Talbot: bitumen (in lithography); tin or, for instance, iodide; lactates and nitrates of silver; carbon processes; uranium nitrates; and chlorides of gold." Parikka, *Geology of Media*, 55.

11. According to Bonneuil and Fressoz, understandings of the Anthropocene enable "the reunion of human (historical) time and Earth (geological) time." Bonneuil and Fressoz, *Shock of the Anthropocene*, 32.

12. Lindsay Jones, editor in chief, *Encyclopedia of Religion*, second edition, volume 6 (Farmington Hills, MI: Thomson Gale, 2005), 3594.

13. Lindsay Jones, editor in chief, *Encyclopedia of Religion*, 3586.

14. Joseph Masco, *The Theater of Operations: National Security Affect from the Cold War to the War on Terror* (Durham: Duke University Press, 2014), 86.

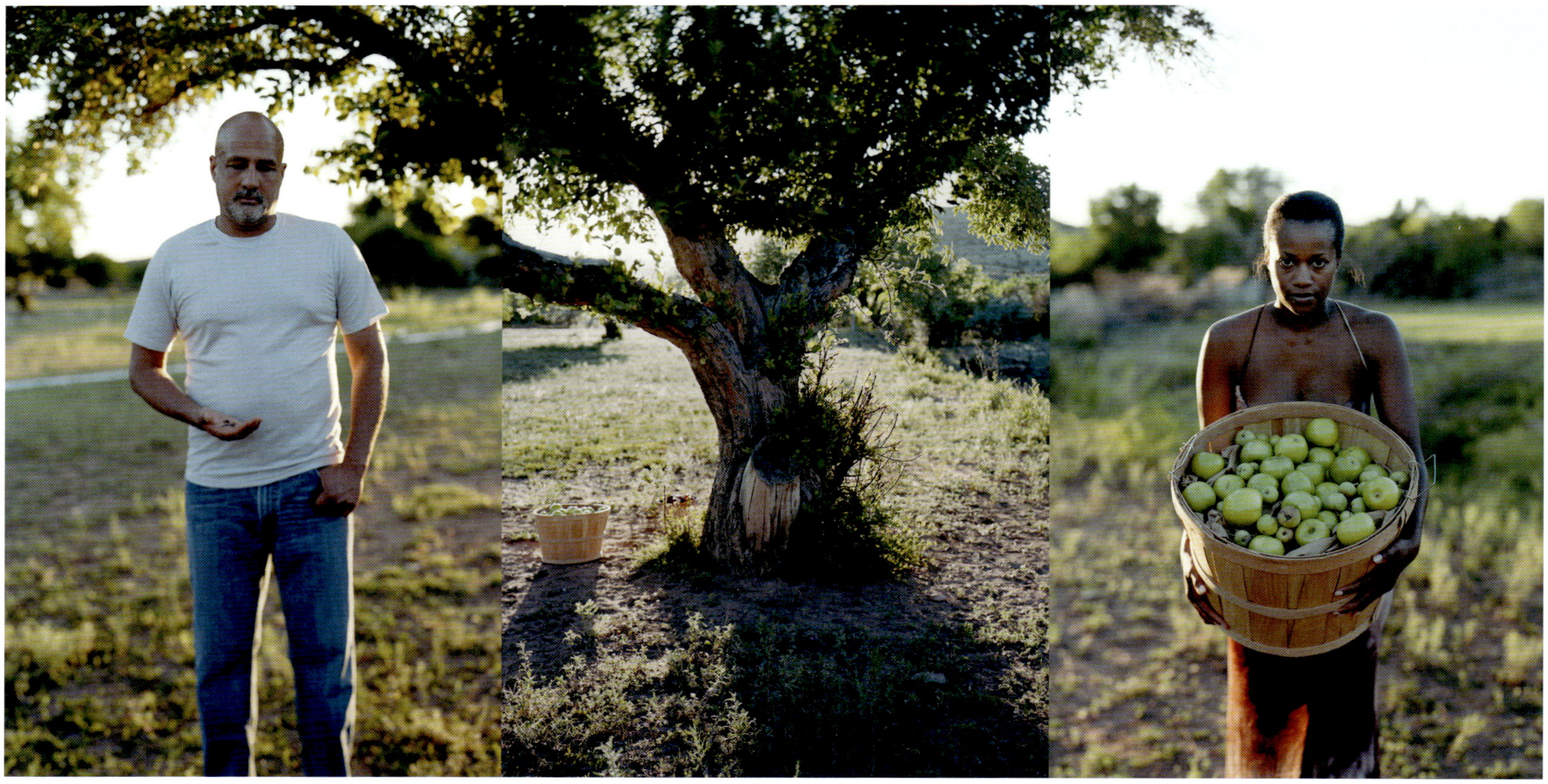

Adam and Eve in La Cienega, New Mexico, 2009–2011

Back to the Garden

Alan Weisman

> We are stardust,
> billion-year-old carbon.
> We are golden,
> caught in the devil's bargain.
> And we've got to get ourselves
> back to the garden.
>
> —Joni Mitchell,
> *Woodstock*, 1969

Everyone knows which garden she meant. Something in us yearns to go back there. But could we, somehow? And to what, exactly, would we be returning?

To anyone whose cultural roots are entwined with a certain mystical history emanating from the Middle East, the Garden of Eden evokes either a carefree time before sin appeared, or a pristine, untainted planet—or possibly both, if you believe that the twenty-first century's eco-perils stem from our sins against nature. But whichever it may be, it hearkens to a time before any of us was alive—yet one for which we oddly feel a visceral nostalgia.

That gut longing is even stranger, given that our notion of Eden comes not from experience, but from a book—in fact, from one of its most problematic passages. The Judeo-Christian Bible begins, logically enough, with God creating the world. Within six days, there are stars overhead, sun by day, moonlight by night, seas and sea creatures, land, plants, land animals, and finally people, both male and female. It's a world we'd basically recognize, other than that it is entirely vegetarian: fruit, seeds, and green herbs shall be for meat, God tells humans and beasts alike.

All of that is described in chapter 1—where, incidentally, there's no mention of any garden: the entire Earth is bountifully blessed. But in chapter 2, after the seventh day when God rests, logic begins to unravel.

Suddenly, the whole story starts over from scratch. This time, there's no workweek plus a Sabbath; instead, God conjures up Heaven and Earth in a single day. And the sequence changes: again, he invents plants early on, but he sets them aside in some celestial nursery because, unlike before, he hasn't created water yet. Next, he does so—then, just as the soil begins to moisten, he selects some dust to mold into a human being. A male.

Only after breathing life up man's nostrils does God add the foliage—but this time, it's to a garden. However gorgeous and fertile it might be, Eden isn't all bliss and leisure. Here, God intends for man to "dress," "keep," and "till the ground." He sets guidelines about what can or can't be harvested, designates a network of rivers for irrigation, and finally, lest man be lonely, creates animals.

Adam gets to name these animals and to identify which are domestic and which are wildlife, yet they fail to assuage his solitude. So, borrowing one of his ribs, God clones him a more tempting companion. And you know the rest.

Much has been made by Biblical scholars of these two conflicting, coexisting creation tales. As their syntax and even their names for God differ (Elohim in chapter 1; Yahweh in chapter 2), it's widely assumed that they were written at different times by different authors, drawing on different versions of earlier Assyrian and Sumerian legends that were later conflated, often awkwardly, into a single testament by anointed redactors, who thrashed out what was canonical and what wasn't.

Small wonder religion breeds so much discord. To literalists who claim that the Bible is divinely inspired and thus entirely true, the second chapter simply details chapter 1's sixth day when man appears, to underscore God's special relationship with us. (Thus, presumably, it's more fitting that humans should precede the flora and fauna over which they hold dominion.)

Babylonian Talmudists tried reconciling another notable discrepancy, over whether woman was born simultaneously with man or later ripped from his rib cage like a bloody afterthought, by declaring that both were true. Appropriating a Sumerian legend of a she-demon, they concluded that chapter 1's female was Adam's rebellious first wife, Lilith, who disdained his authority and abandoned him. In chapter 2, God tries to ensure that this doesn't happen again. The woman fashioned from Adam's own bone is merely his "help meet"—so subservient she doesn't even get a proper name until after she also proves too disobedient and seductive to control and gets them booted from the garden.

And us, too, so the tale goes—but that may well be why both creation myths are included in Genesis. Taken together, they reflect an evolutionary truth sensed by whoever compiled the Bible from collective memories inherited through their forebears' stories and perhaps even through their cells and genes. In one version, we had the whole Earth's bounty to pluck for our nourishment. (Archeology, littered with spear points, choppers, and blades, belies the vegetarianism, which probably seeped into the account much later, during the time of Isaiah: an ardent herbivore who condemned animal sacrifice and who prophesized a post-Messianic paradise where wolves would dwell among lambs and leopards lie peacefully with kids.)

But in version two, we are no longer hunter-gatherers. We're stewards of orchards, tillers of soil, and pastoralists of the cattle that Adam named. Thus was born agriculture. Yet by giving us a garden, was the Lord setting us up for a Fall?

To grow, conveniently close at hand, food that we once roamed hither and yon to scrounge might have seemed to our ancestors an astonishing improvement over what chapter 1's God—or gods, or Mother Nature—had originally given us. But since a few farmers and herdsmen can feed many, the concentration of sustenance eventually led to concentrations of people. Soon our increasing numbers had to seek more fields and pastures, but they found that the greater world wasn't all an oasis like Eden. Before long, as Genesis recounts repeatedly, people were warring over the choicest lands, and especially over water. This has not abated.

What Genesis doesn't mention, of course, is that we were fighting long before we were human: Our chimpanzee cousins have never traded foraging in trees for working the soil, but they battle for territory anyhow, as our common primate ancestors also surely did. What sets us apart from them, however, is what else agriculture ignited.

Once farming and grazing were established, for the first time in our or any species' history, a significant portion of the population didn't have to spend all day finding food. Gradually people devised new things to do for a living: from artisanship to architecture to trade, from prostitution to priesthood. Soon many of us lived farther from crops and livestock and nearer to each other, in settlements, then villages, then cities. Whatever grief early urban dwellers might have sustained for some fabled, lost Edenic profusion gave way to the thrill of something entirely new: civilization.

Long before that happened, successive waves of *Homo sapiens* had begun dispersing from our African origins. By fifty thousand years ago, we had spread all the way to Australia—and, by about fourteen thousand years ago, to the Americas. As a result, civilization emerged independently in several places. Among the first to appear was likely the one in the fertile crescent that our prehistoric ancestors followed from Africa into Asia. As we continue to learn about it, it turns out that Eden may be more than just a metaphor.

Genesis locates the Garden of Eden quite precisely, near the confluence of four rivers: the Tigris and Euphrates—whose joint Mesopotamian valley is often called civilization's cradle—and the Gihon and Pison. The 1611 English-language King James Bible placed the Gihon across the Red Sea in Ethiopia, rather far from Mesopotamia. But several modern scholars believe that its translators confused Ethiopia's ancient name,

Cush, with a similar Hebrew name for the land of the Kashshites, who lived in what is now Iran's Zagros Mountains. From there a substantial river, the Karun, flows southward, reaching a delta just above the Persian Gulf at a point almost directly opposite Saudi Arabia's north-flowing Wadi Al-Batin.

That point is sixty miles below where the Tigris and Euphrates meet today, but back then—as noted by archaeologists who contend that the Karun and Wadi Al-Batin are the Bible's Gihon and Pison—local topography was far different.

Until about eleven thousand years ago, repeated Pleistocene glaciations locked much of the world's water into ice sheets, and oceans were hundreds of feet lower than today. When the first *Homo sapiens* reached the Persian Gulf, its upper third was well above sea level. British archaeologist Jeffrey Rose calls it the Gulf Oasis: a luxuriant delta, continuously fed by nutrient-rich sediments borne by the Tigris, Euphrates, Karun, and Wadi Al-Batin.

For hunter-gatherers, it would have been a paradise. But post-Pleistocene warming, which allowed their descendants to learn how to farm local wild wheat and legumes, also meant that to the north, vast glaciers were melting. As the thawing accelerated, worldwide coastal flooding commenced. At its peak, eight thousand years ago, Persian Gulf waters were encroaching so fast, according to American archaeologist and Middle East specialist Juris Zarins, that with each generation they advanced several kilometers farther inland. By the time civilization awakened in its cradle a millennium later, paradise had been lost beneath rising seas.

Fall and flood: memories passed from prehistoric ancestors to Mesopotamia's Sumerians, Assyrians, Akkadians, and Babylonians, who passed them on to the Hebrews, and to us. Today, the last traces we have of the oasis that inspired them are some lush wetlands in Iraq's Tigris-Euphrates floodplain, upstream of that now-submerged Eden. During the 1990s these rare marshes were also nearly lost, when Iraqi dictator Saddam Hussein tried to drain them to expel Shia rebels. But following his ousting, a courageous local and international effort has recovered almost half of them.

Staunching that ecological disaster is one of the few encouraging stories from that tortured region in recent years—but unless we act decisively, it could prove in vain, as ice is melting again. In an elaborate, circuitous lineage that humanity has confected, agriculture begat civilization, which begat industrialization, which inevitably—some might say incestuously—begat industrialized agriculture. Force-feeding soil with chemical fertilizer, and genetically developing hybrid grains with many extra kernels per stalk, we multiplied our foodstuffs far beyond what nature ever could supply. But unintentionally this unprecedented abundance also multiplied ourselves.

Averting famine means more people live to beget more people. And so human numbers quadrupled during the past century—the most abnormal population leap for any large species in the history of biology. With all of us hooked on energy, the exhaust from our swollen industrial civilization keeps turning up the heat, forcing seawater up river deltas and over coastlines. If the result of Adam and Eve's original sin was procreation, here we are again, headed for another fall.

Pushed to this modern brink, more than ever we long to reclaim the fresh hope of Eden. Yet lest we despair, it's comforting to realize that one day this Earth will surely see Eden restored.

Should our yearning help us muster the environmental wisdom and the will to act upon it, we'll do whatever it takes to bring humans back into balance with the rest of nature.

Or, should we fail, we'll drag many more species and much priceless earthly beauty down with us. But not forever. This planet has endured enormous losses and massive extinctions before, from cataclysmic volcanic eruptions to colliding asteroids—yet each time, life, miraculously resilient, has rebounded, lovelier than ever. As it will again.

One way or another, we'll be there too. Either we'll have survived our excesses, and, like Noah, saved enough animals to give our own species another chance—or, like Adam, we will have returned to the dust from which we sprang.

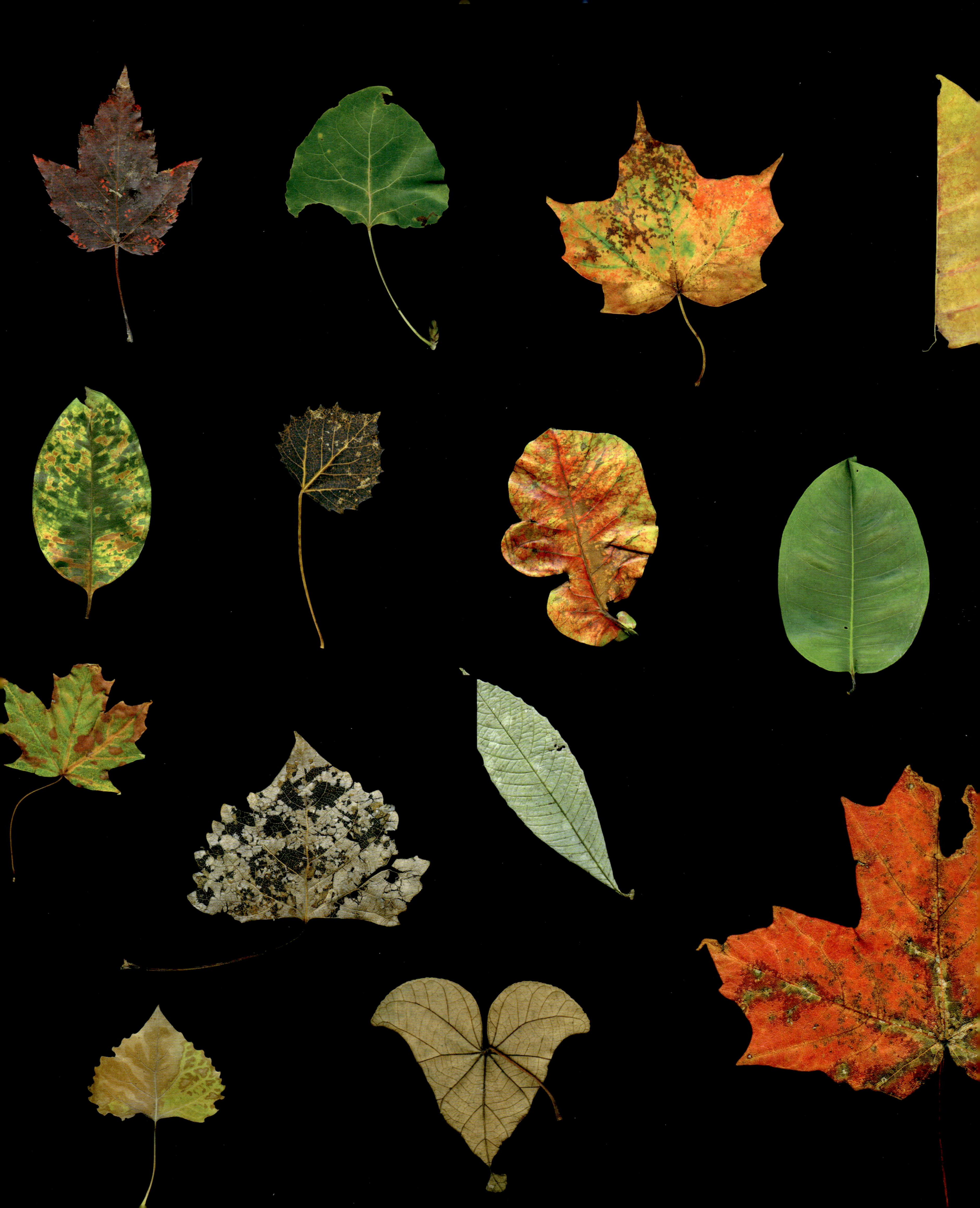

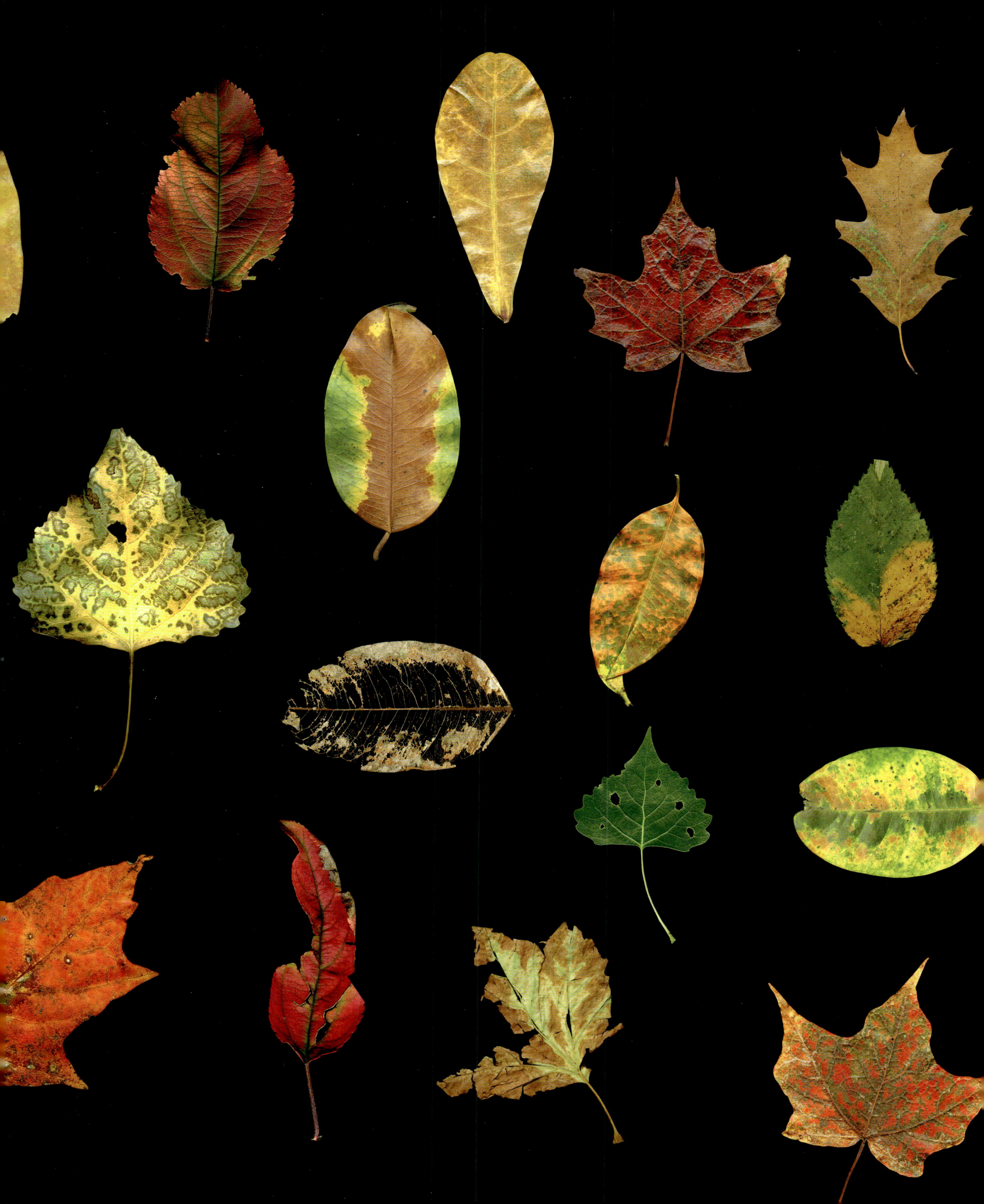

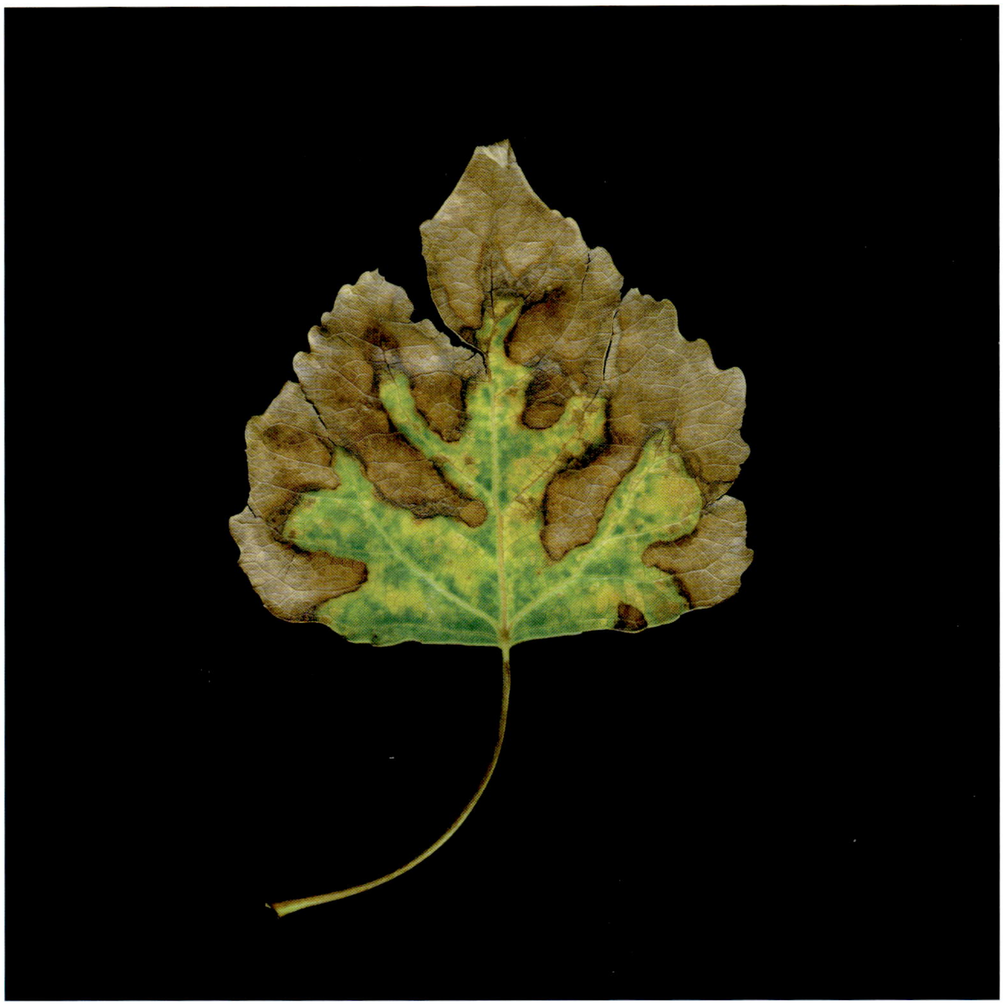

Photosynthesis leaf #2, 2012, 17 × 17 in.

Photosynthesis

Somewhere between our own culturally specific creation myths and the Big Bang, our planet was born.

Out of nothing, into something; out of darkness, a first breath.
From tightly compacted space to infinite spaciousness.
Out of nothing, something.

For eons, the ocean of our atmosphere has cradled us. But now, as it thins from carbon emissions and hydro-fluorocarbons, it may not be able to provide life support to our earth. Upon this planet, lots of things are upside down. Photography allows particles of light to illuminate our present reality, upside down and backward, onto our retina, a piece of film, or a digital light-recording system. It's the artist's job to turn things right-side up in a different way.

Photosynthesis is the predominant light-recording system that sustains life on earth. Sunlight converts Co2 into sugars, releasing energy and nourishment mainly for plant sources known as *phototrophs*. A by-product of this process is *respiration* whereby plant sources exhale oxygen that we *anthropoids* inhale while exhaling Co2 for plants in return. Is it possible to restore any balance between oxygen and carbon, between we humans and the natural world? Or are we prisoners of global greed that is robbing us of our Edenic right to clean air, soil, and water?

This Post-Edenic symbiotic exchange is the subject of this work.

All works in this section are pigment prints on 100 percent rag watercolor paper

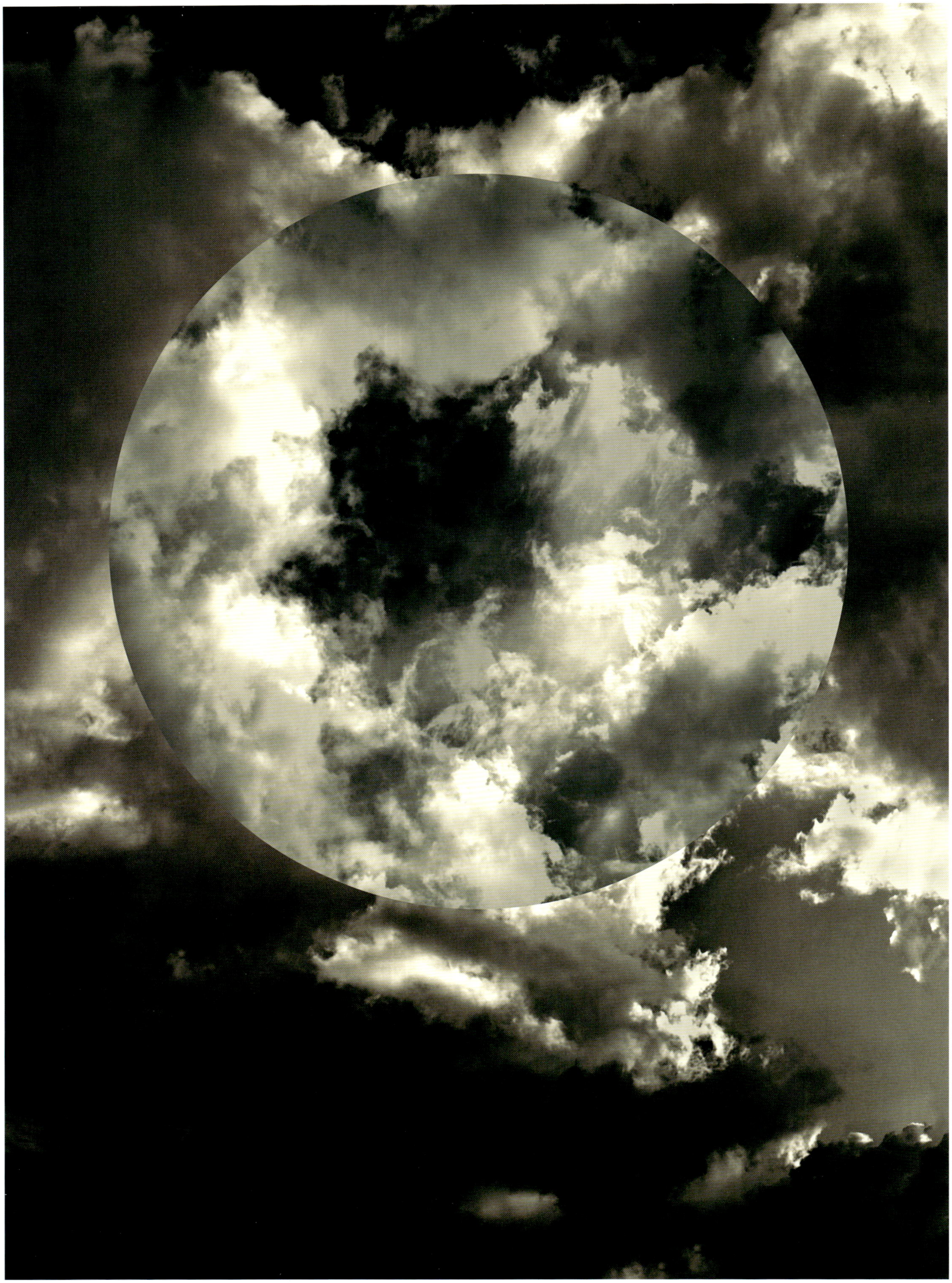

Winter Cloud (The Ocean of the *Atmosphere*), 2009–2011, 46 × 33 in.

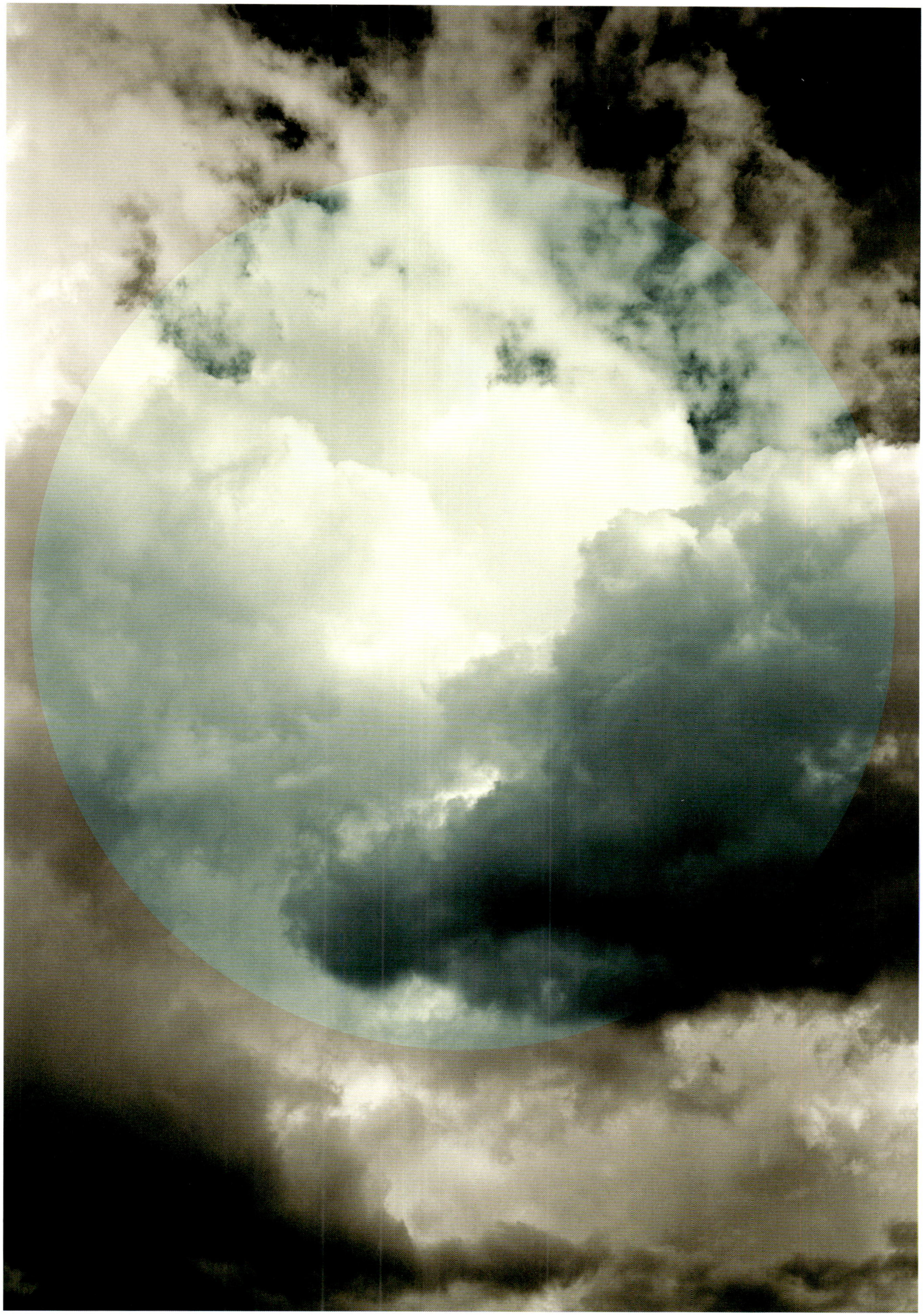

Gaia Cloud, 2009–2011, 43 × 30 ¼ in.

Winter Seasonal, Vermont, 2009–2011, 52 ¼ × 40 in.

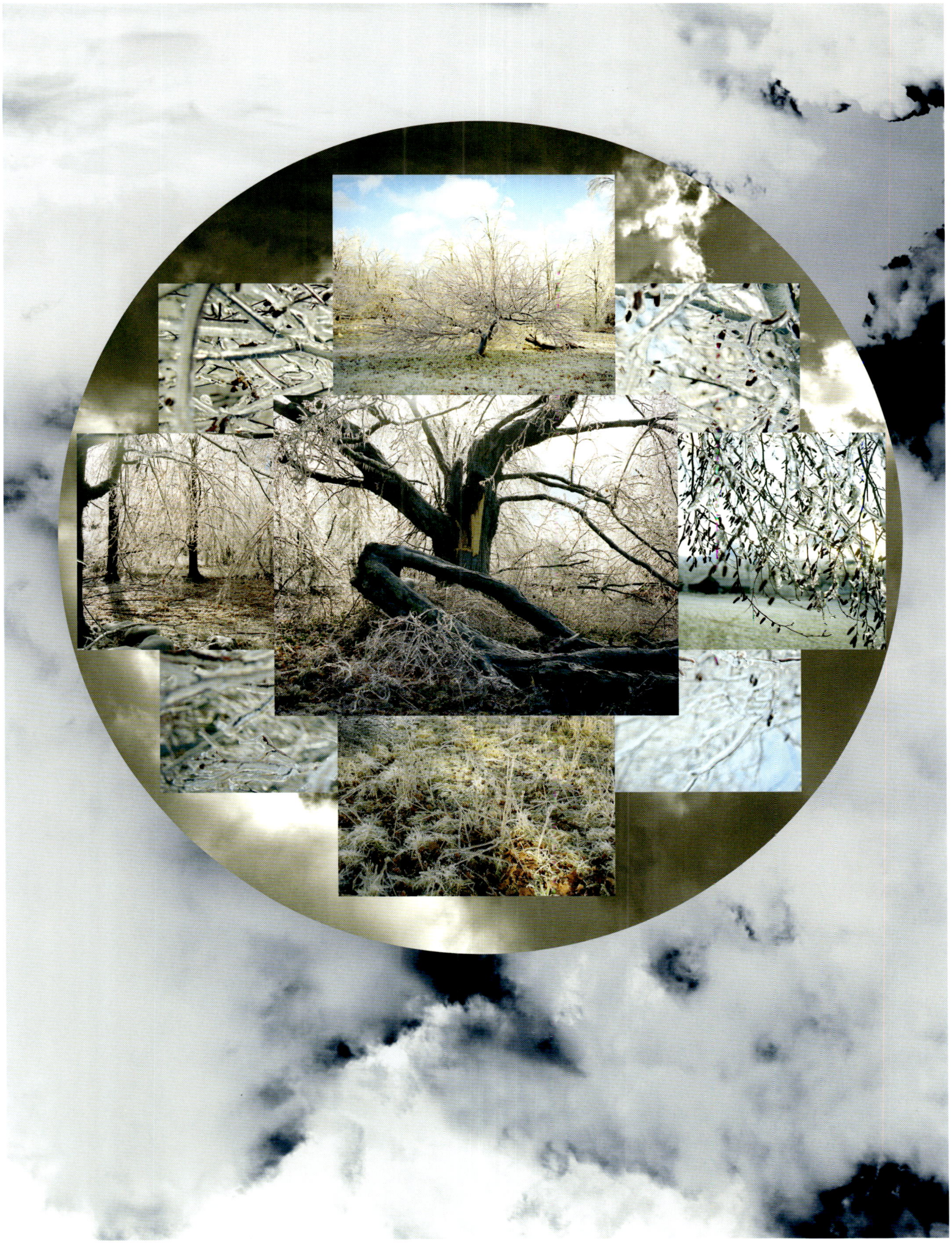

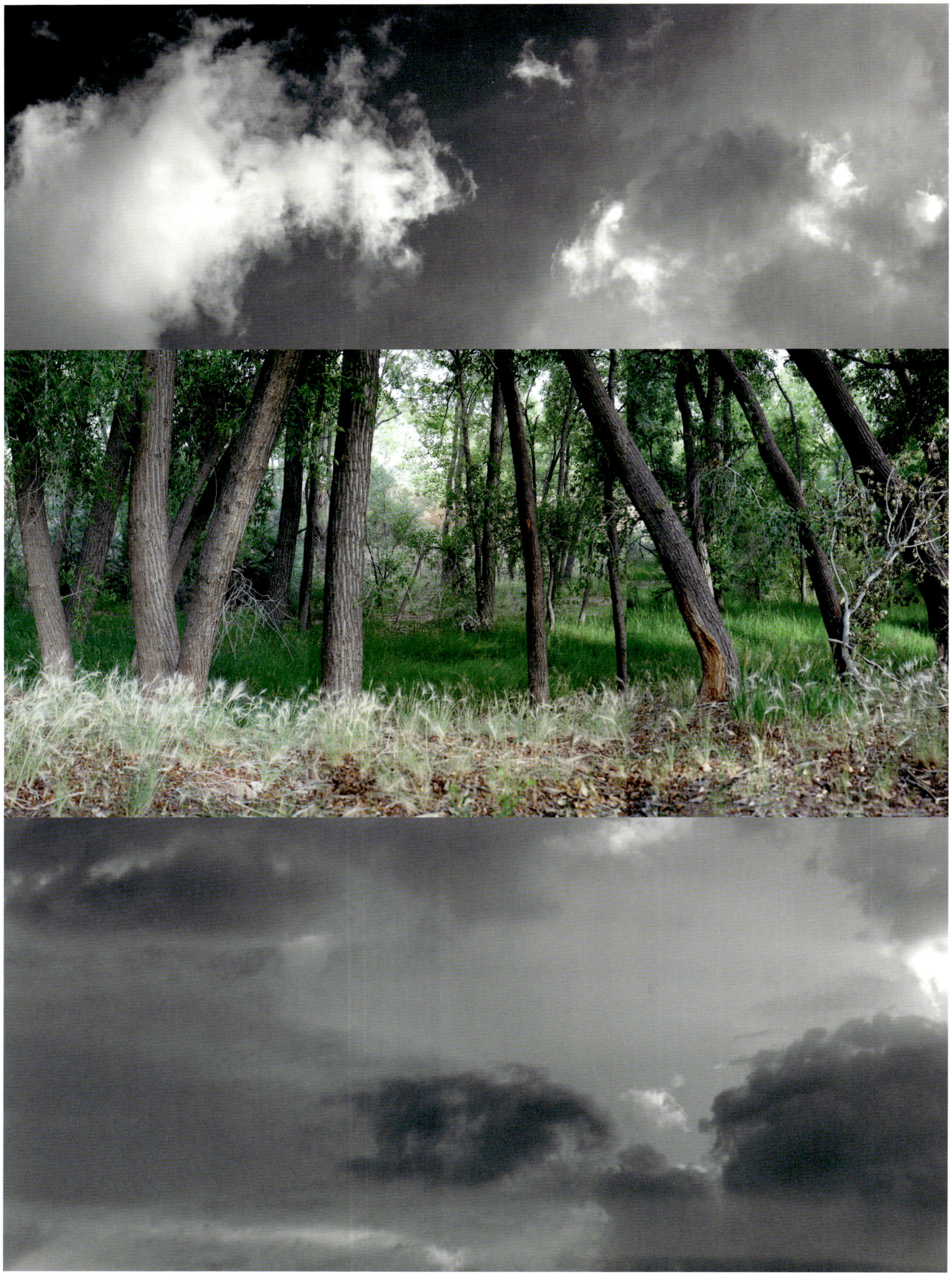

Summer Seasonal, New Mexico, 2009–2011, 54 × 40 in.

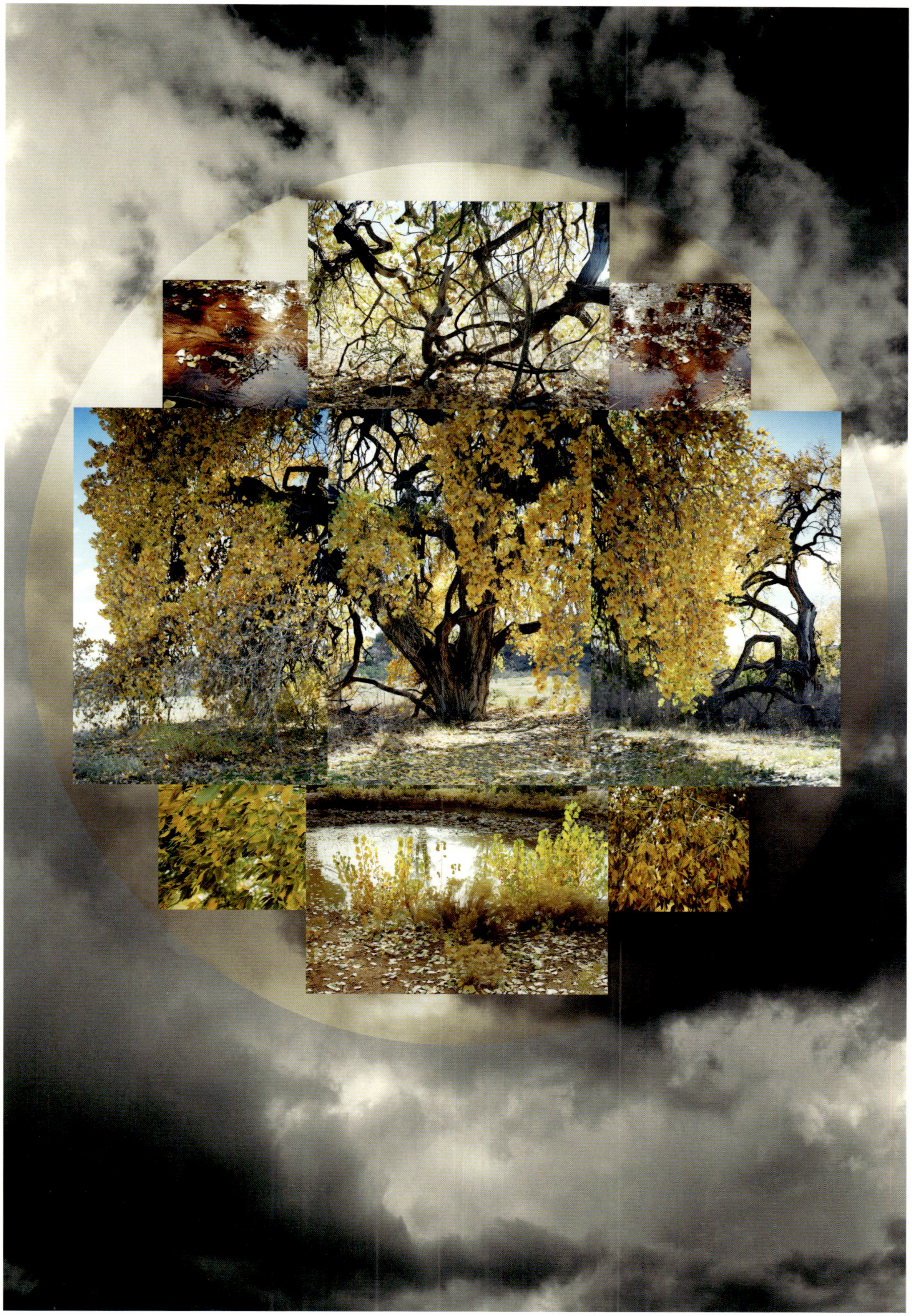

Fall Seasonal, 2009–2011, New Mexico, 57 ½ × 40 in.

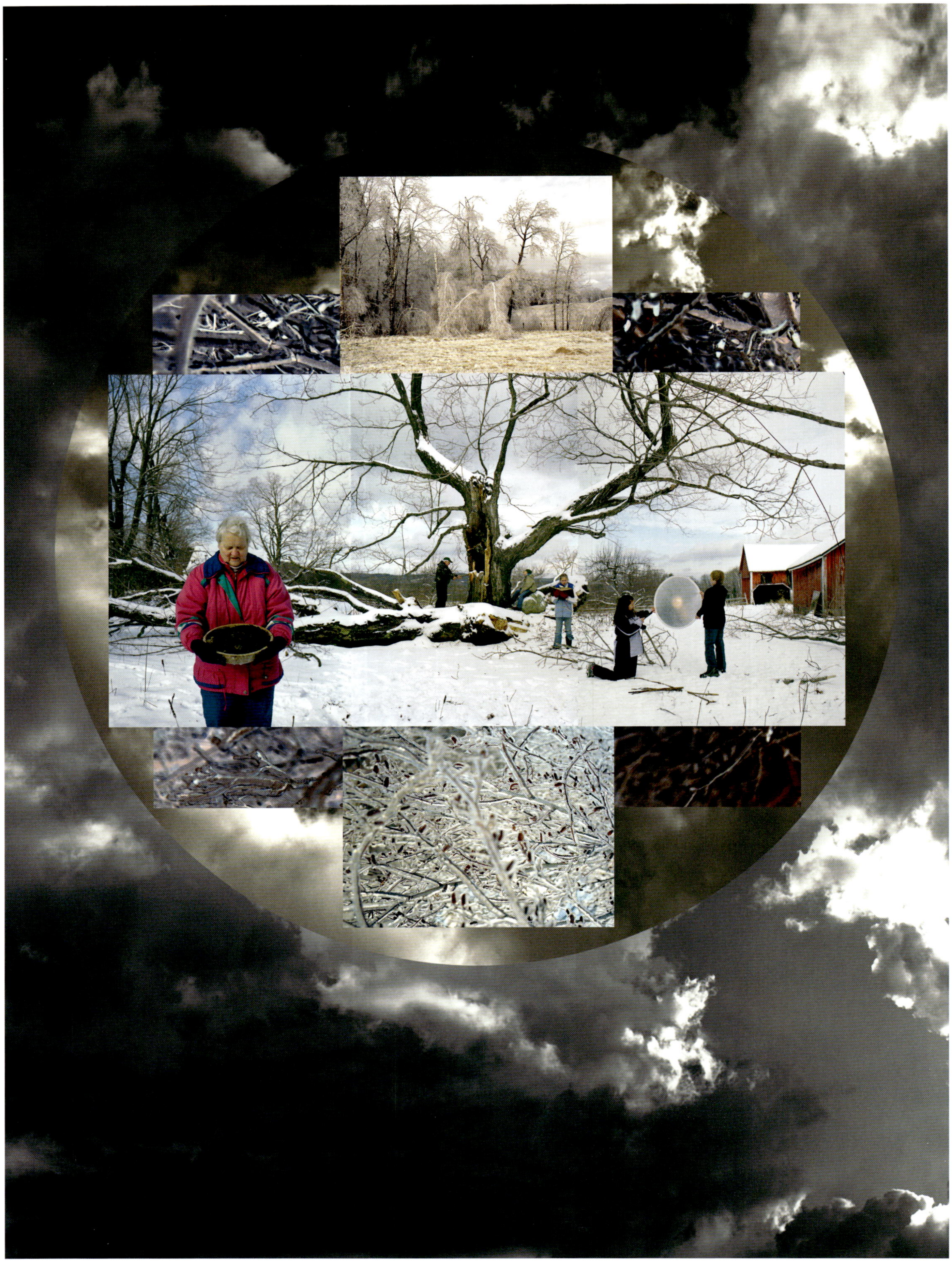

Winter Solstice, 2009–2011, Vermont, 53 × 40 in.

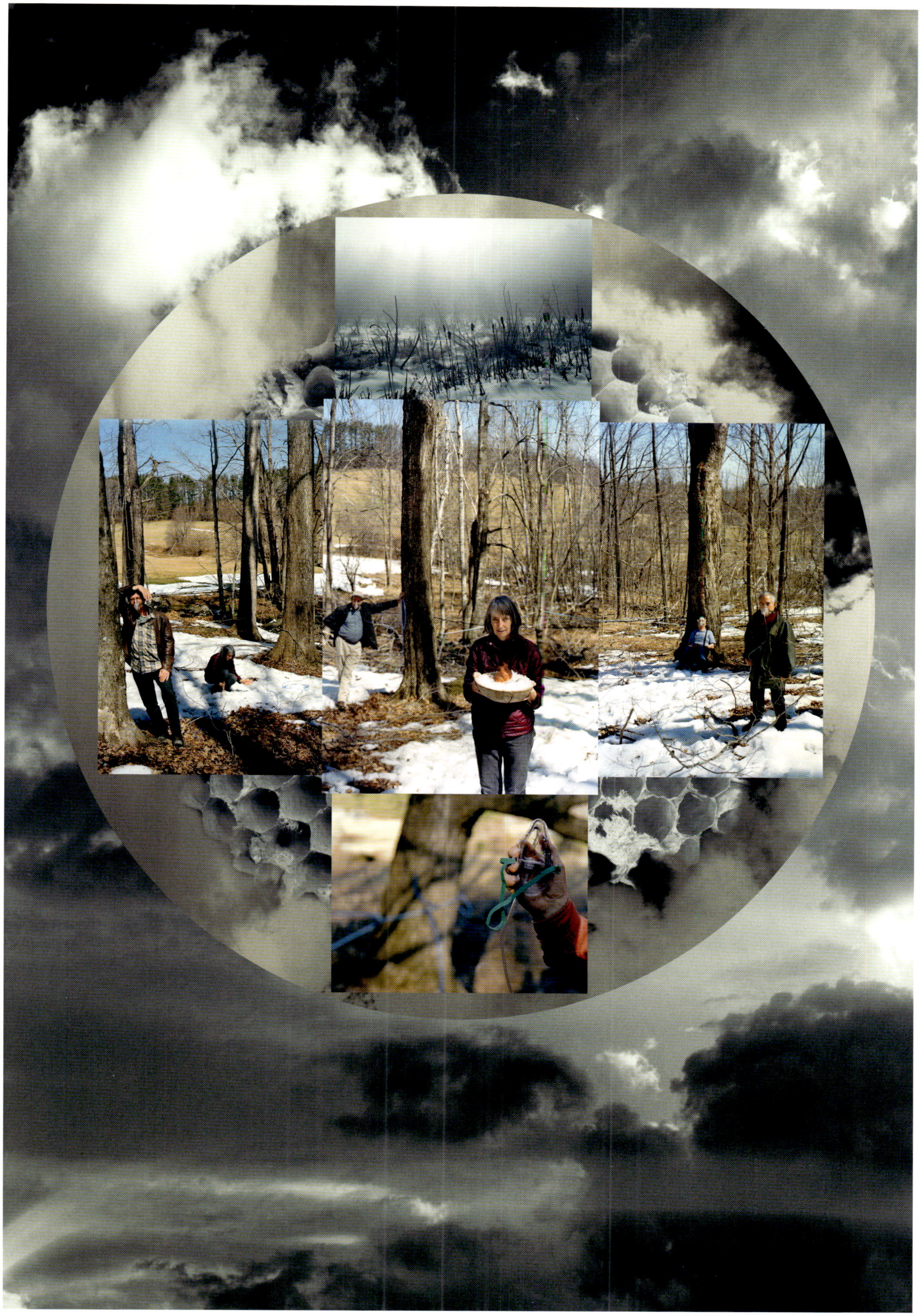

Spring Equinox, 2009–2011, Vermont, 57 × 40 in.

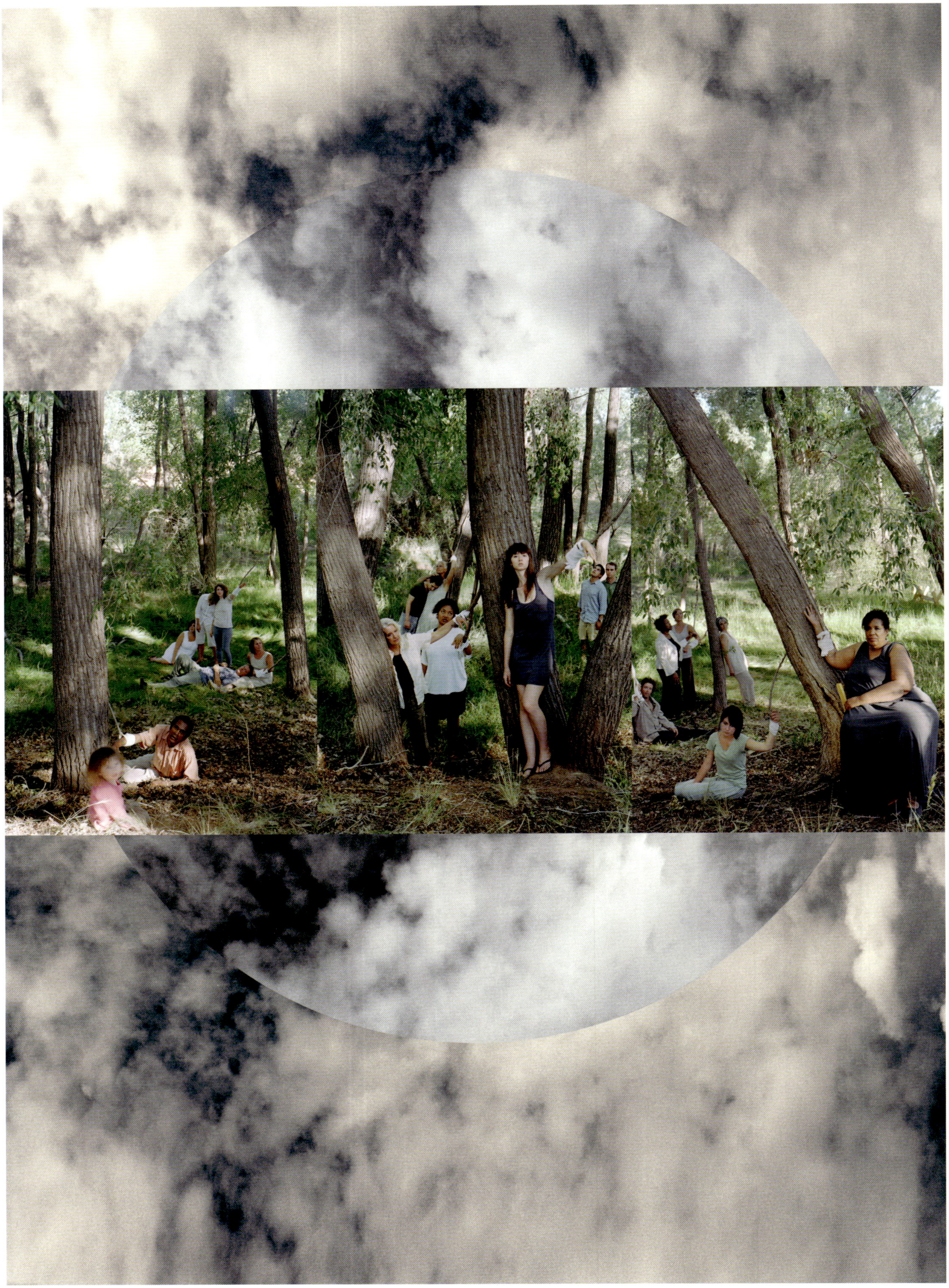

Summer Solstice, 2009–2011, New Mexico, 54 × 40 in.

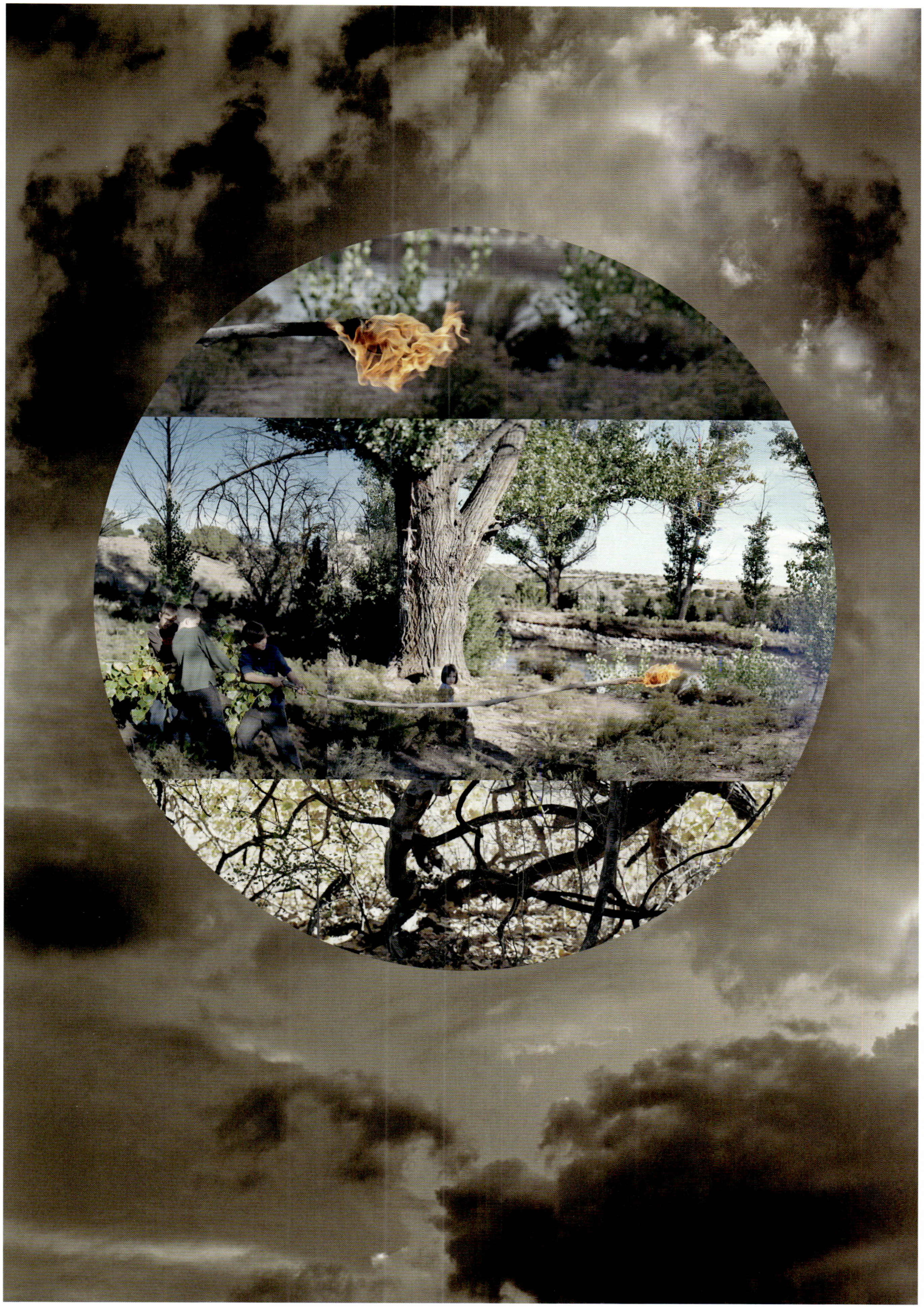

Fall Equinox, 2009–2011, New Mexico, 56 ½ × 40 in.

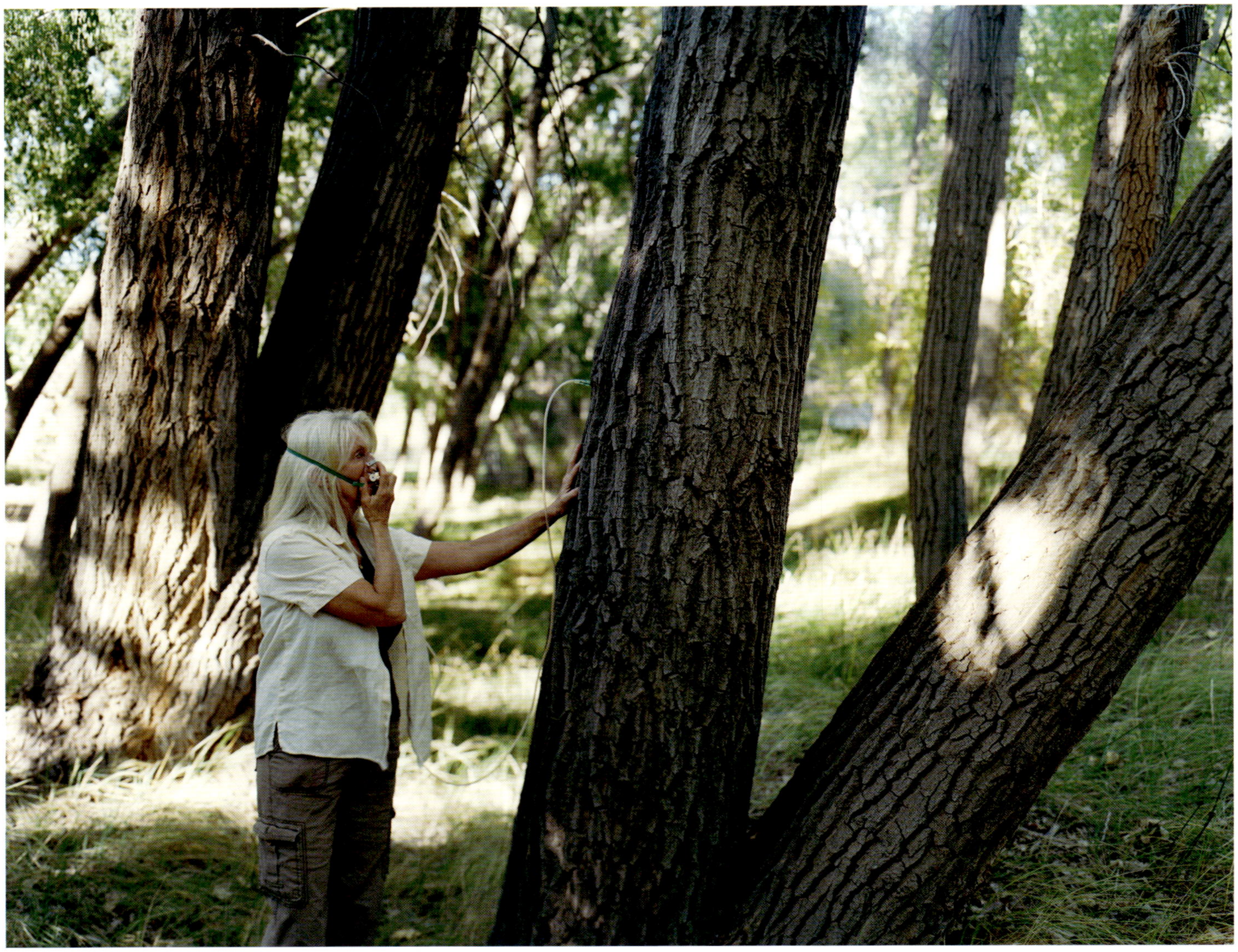

Respiration, 2009–2011, New Mexico, 33 ¾ × 44 in.

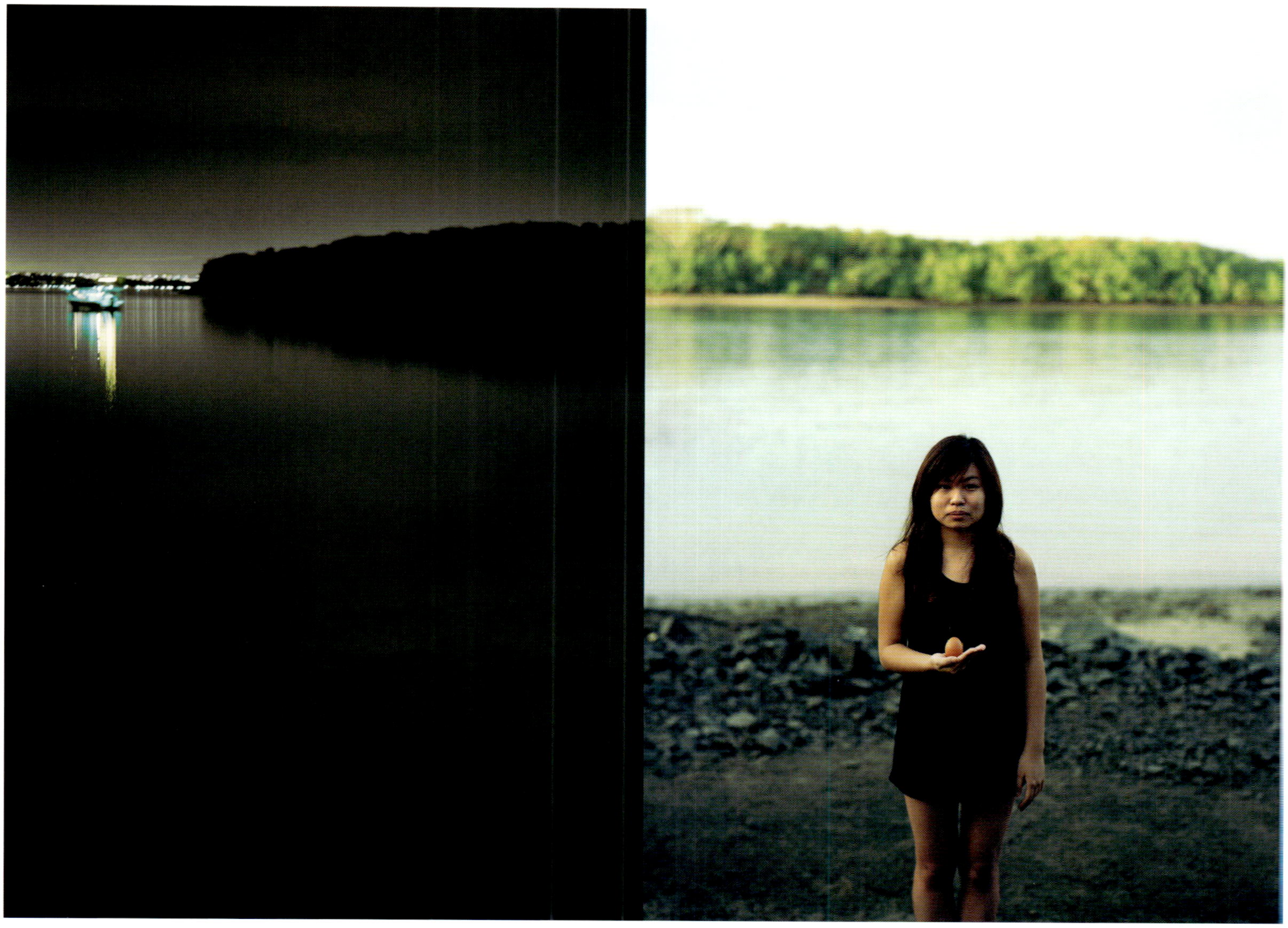

Spring Equinox #2, Pulau Ubin, Singapore, 2010, 34 × 47 ¾ in.

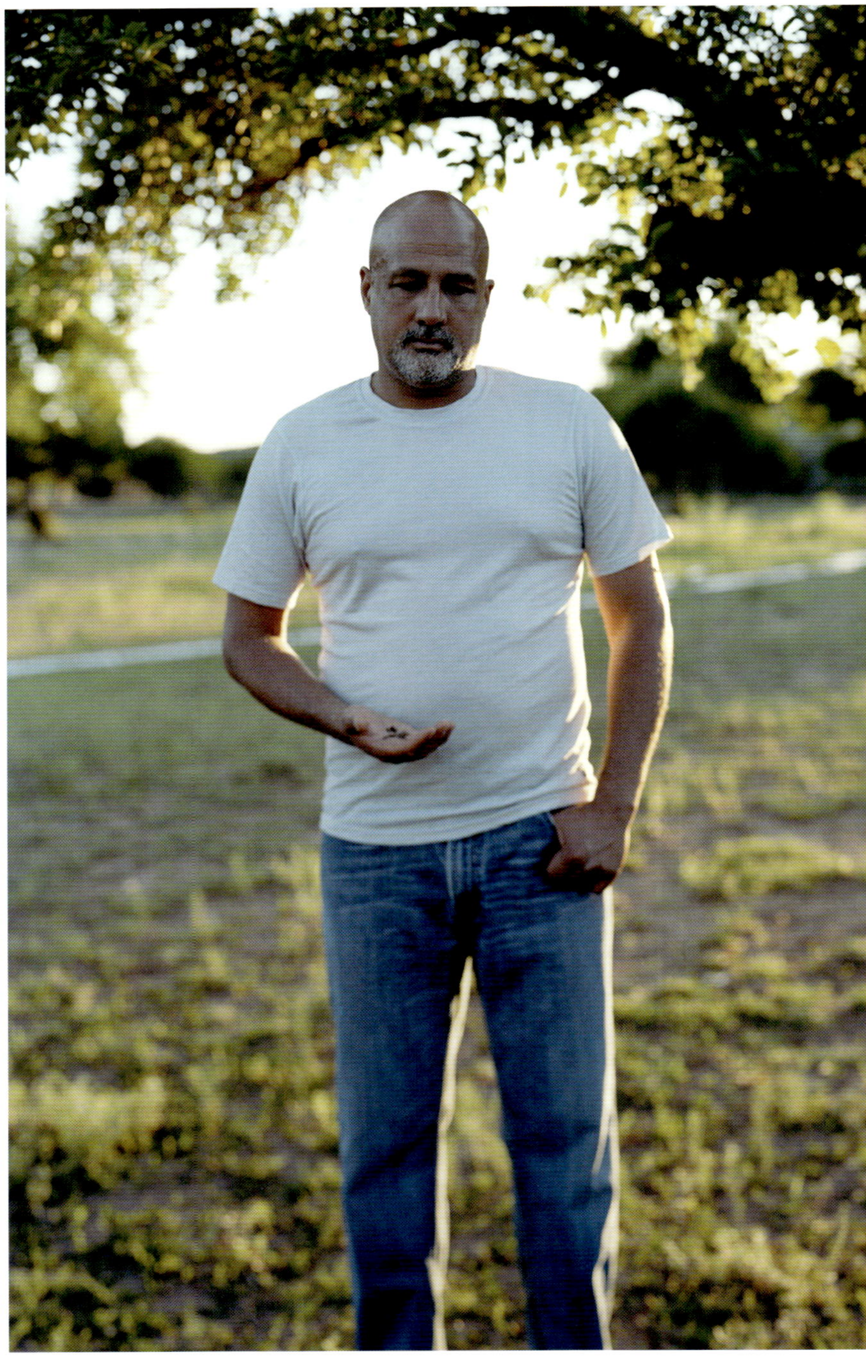

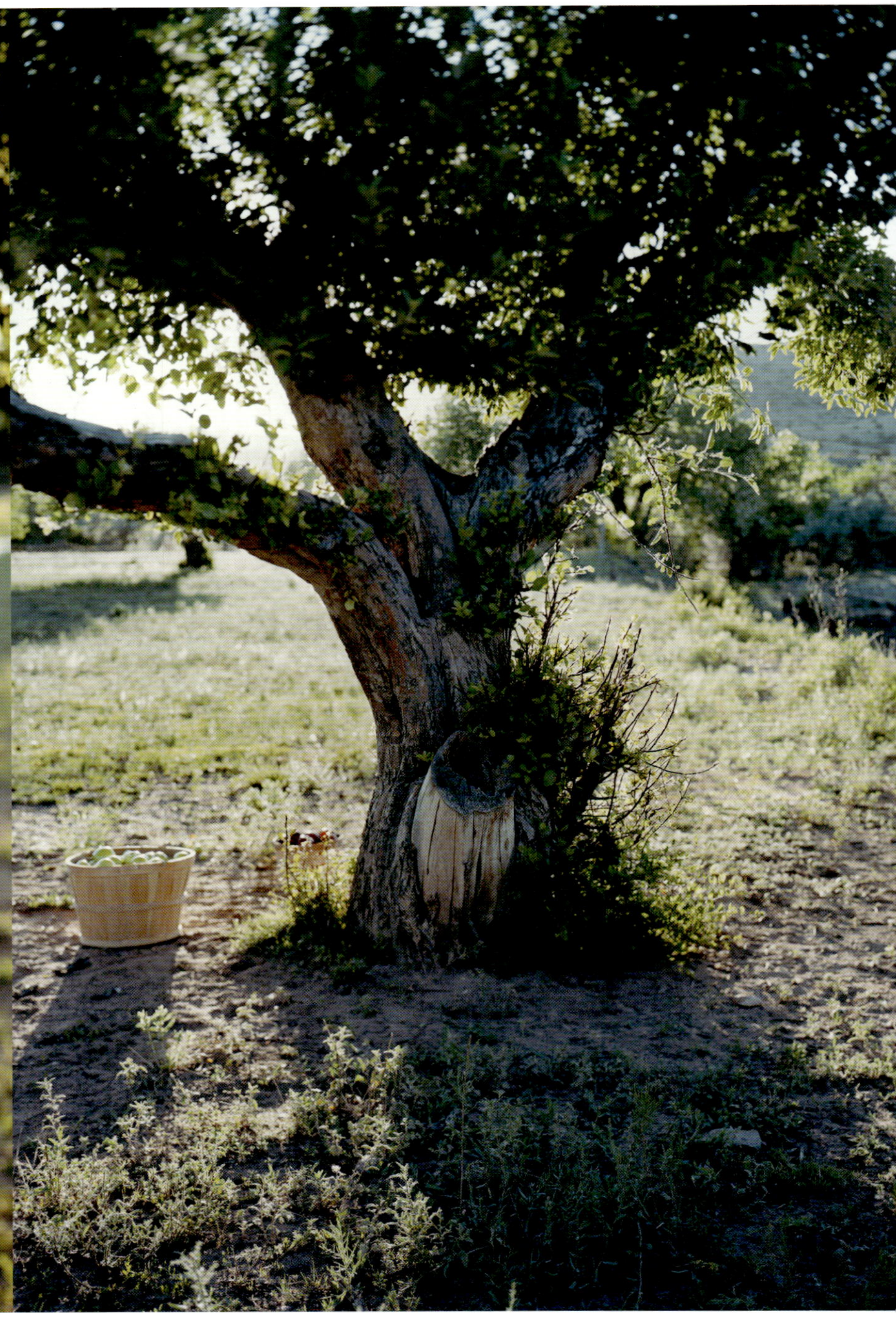

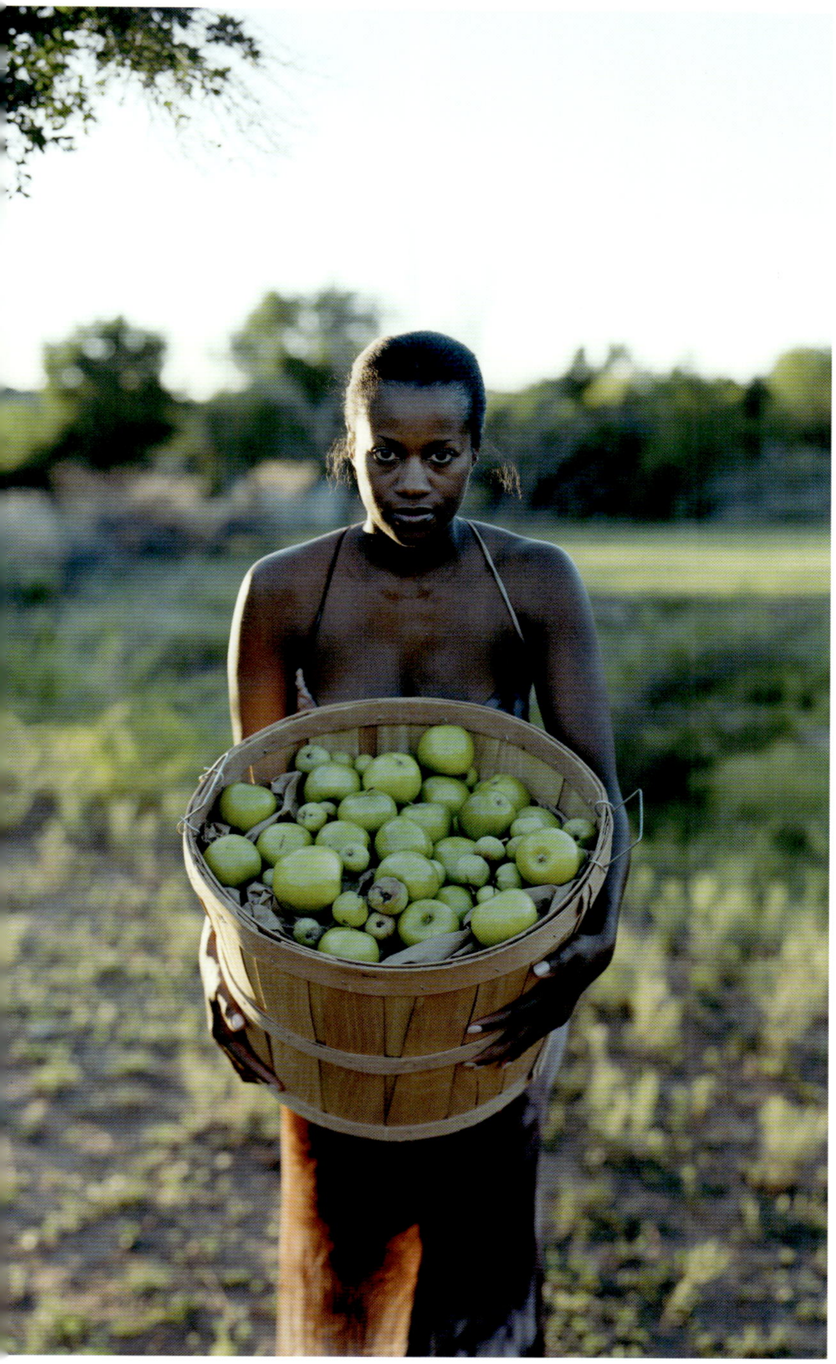

Adam and Eve in La Cienega, New Mexico, 2009–2011, 32 ½ × 64 ½ in.

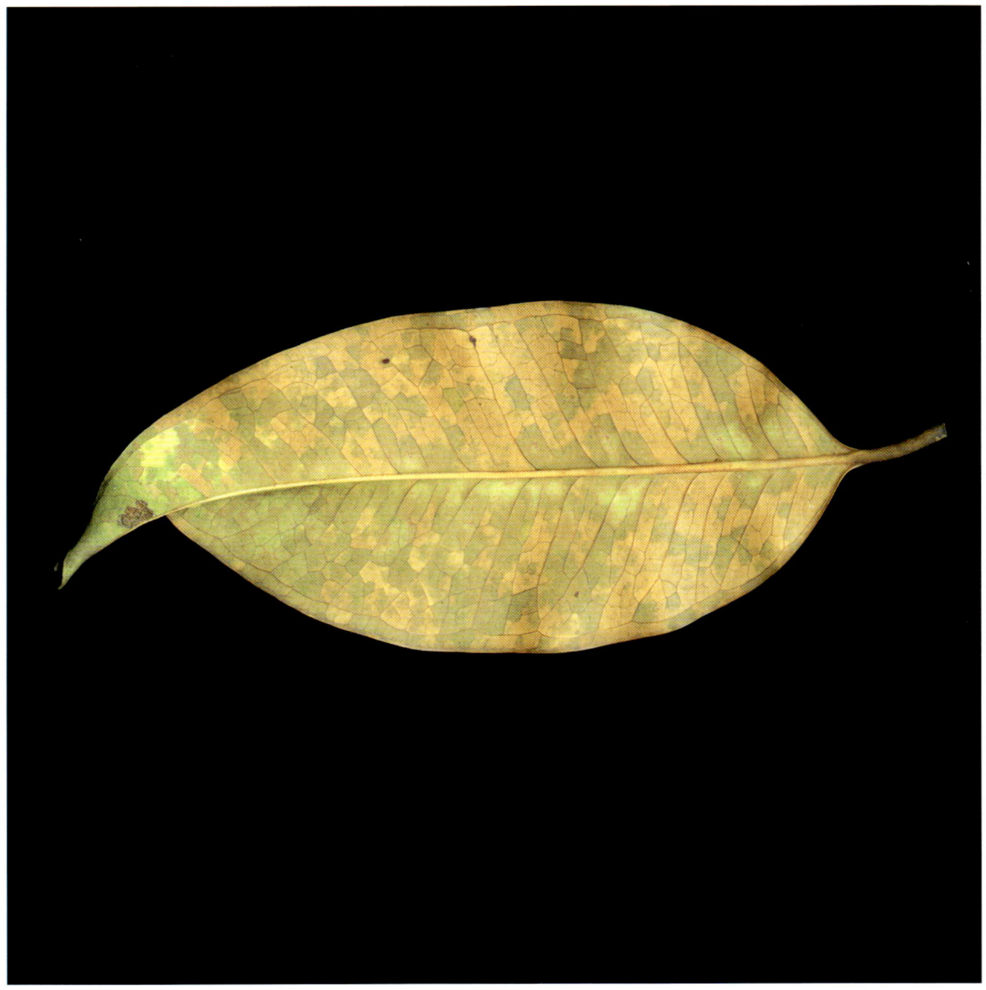

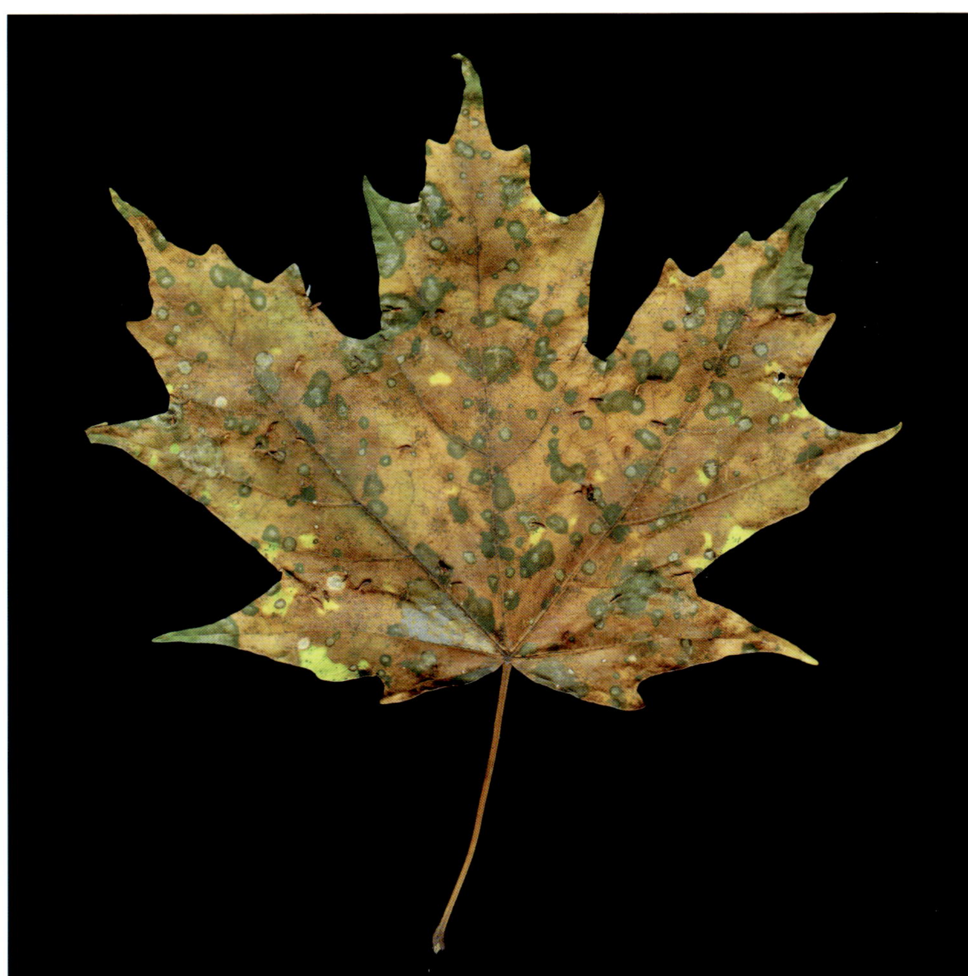

Photosynthesis Leaf Diptych, 2012,
38 ½ × 18 ¾ in.

Tembusa Leaf, Singpaore

Vermont Spotted Maple

Vermont Birch Leaf from Behind, 2012, 17 × 17 in.

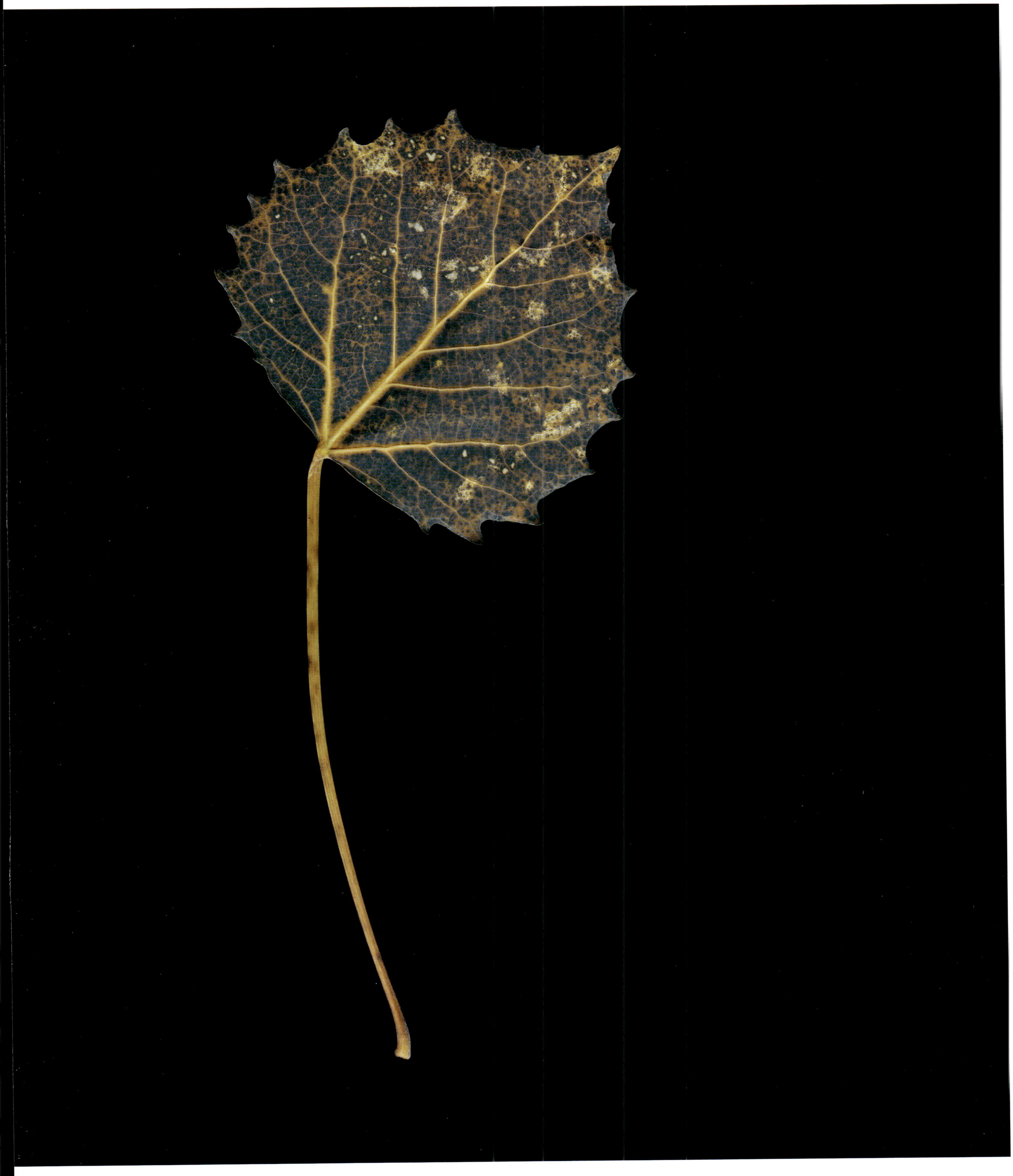

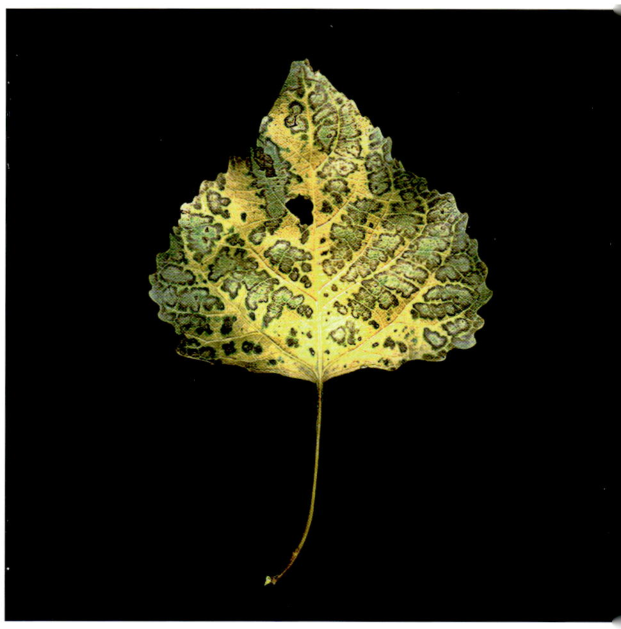

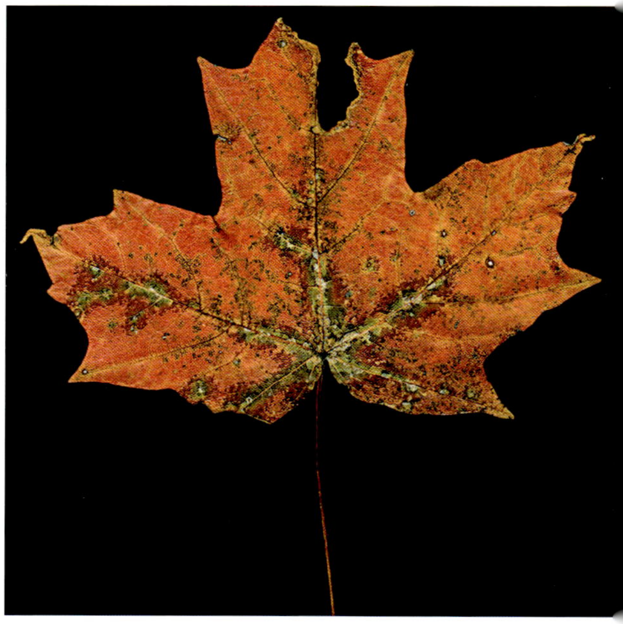

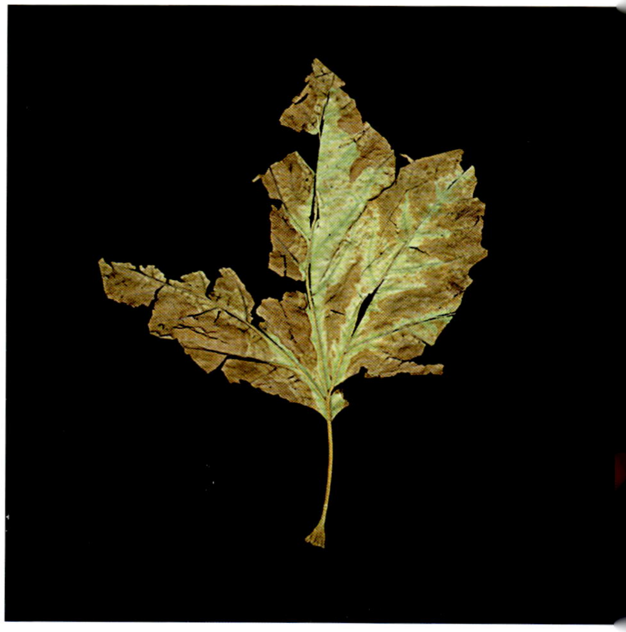

Photosynthesis Leaf Grid, 2009–2012, 9 frames, 18 ¾ × 18 ¾ in. (prints 17 × 17 in.), total area 58 ¼ × 58 ¼ in.

Photosynthesis leaf #1

Equilateral Leaf, Singapore

Pulua Ubin half leaf, Singapore

Red Dinosaur Maple, Vermont

Winter Filagree, Cottonwood, New Mexico

Crabapple Iridescent, New Mexico

Albuquerque leaf, New Mexico

Angsana leaf, Singapore

Fall Palette Maple, Vermont

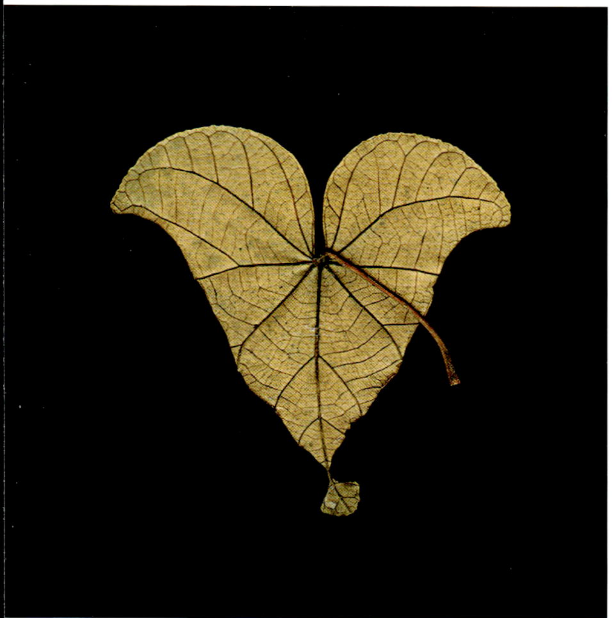
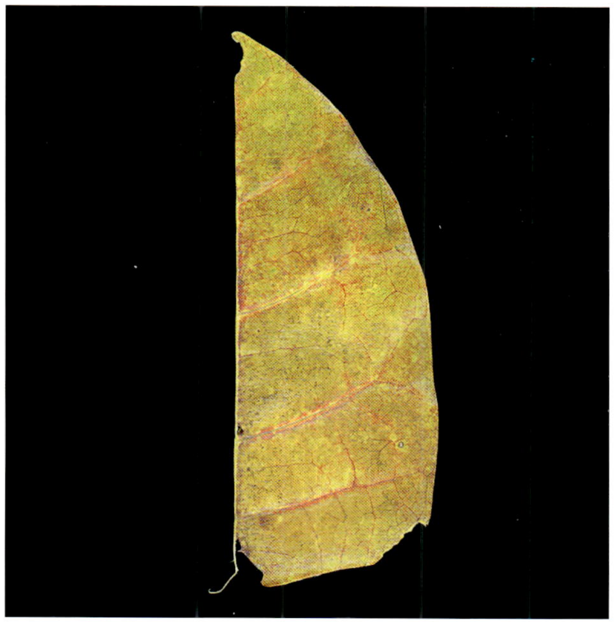
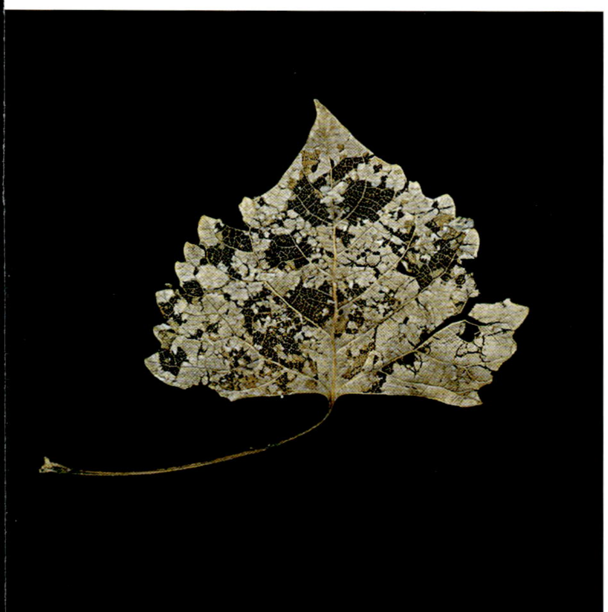
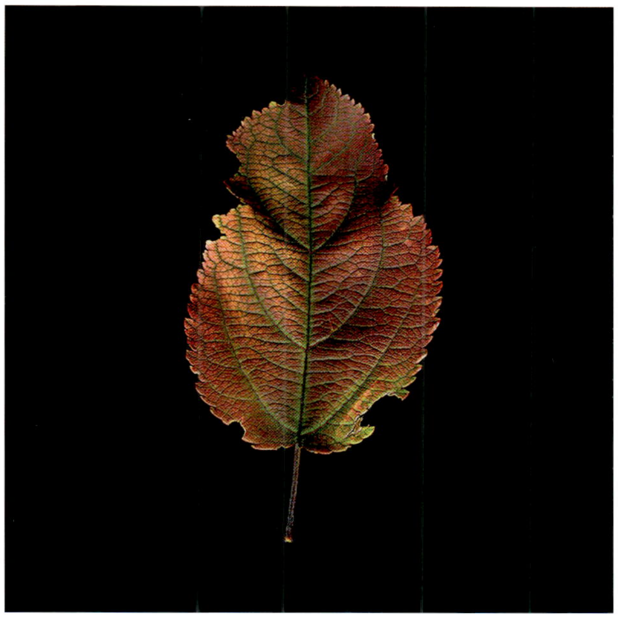
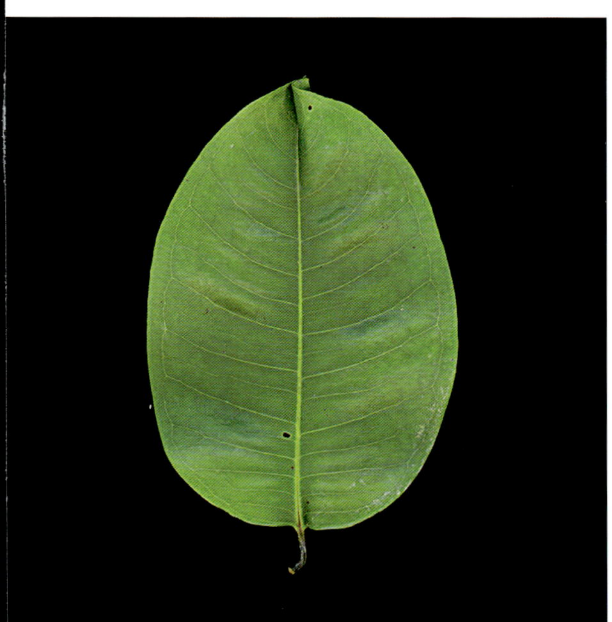
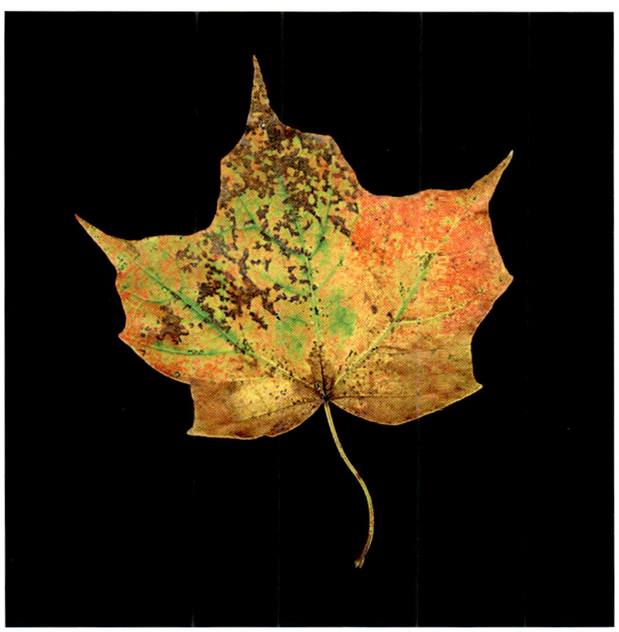

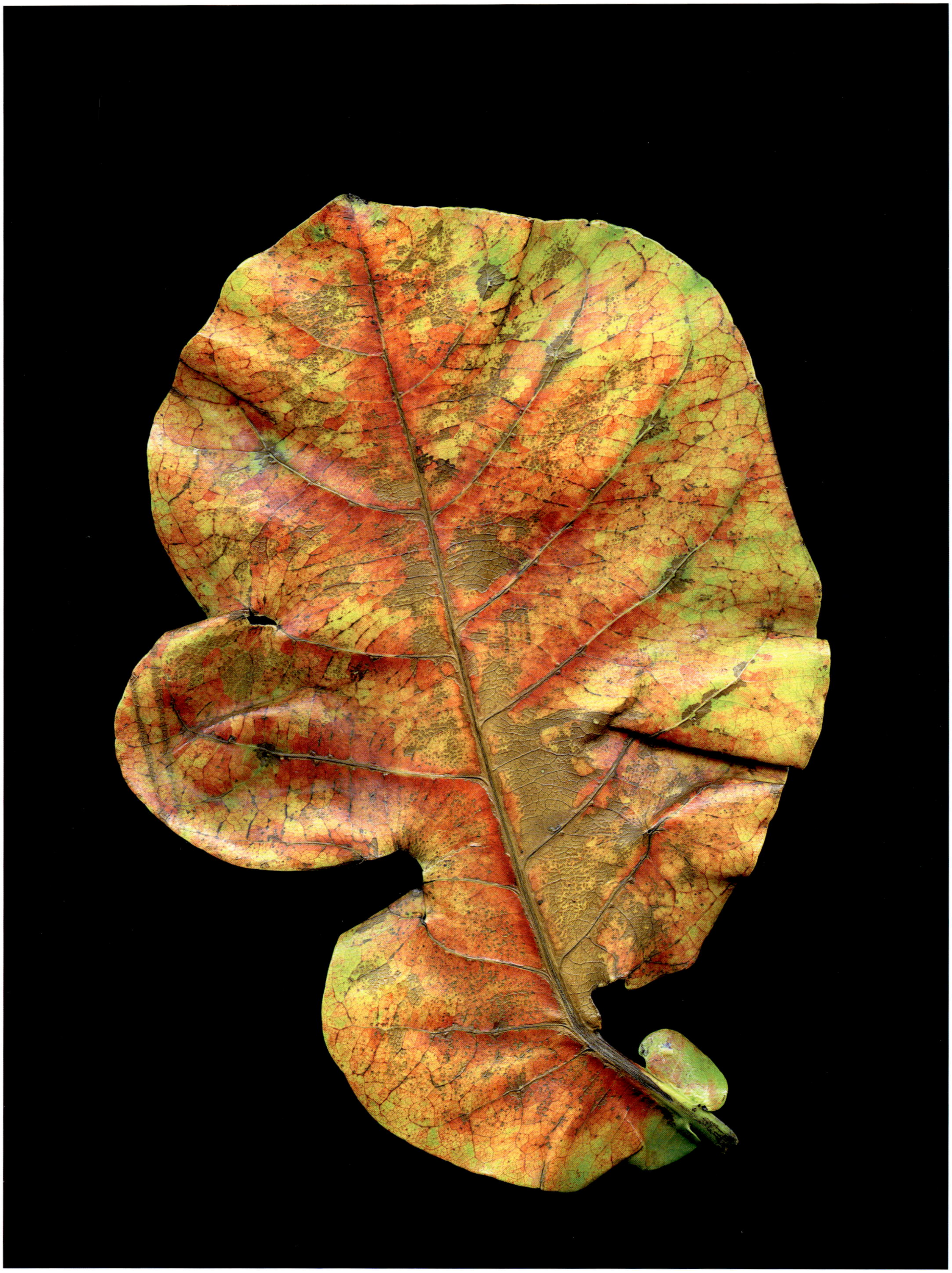

Sibu Island leaf, Malaysia 2015

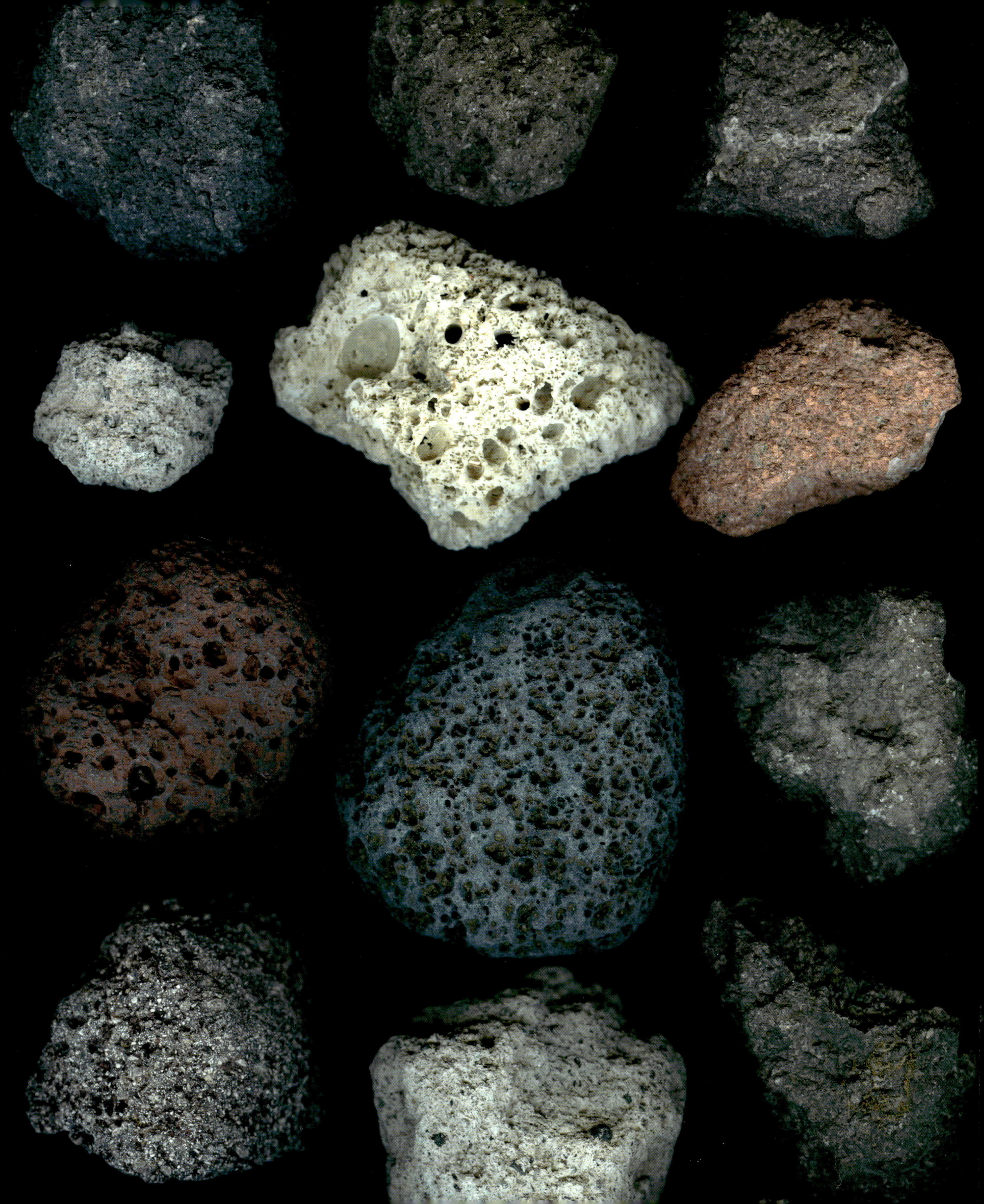

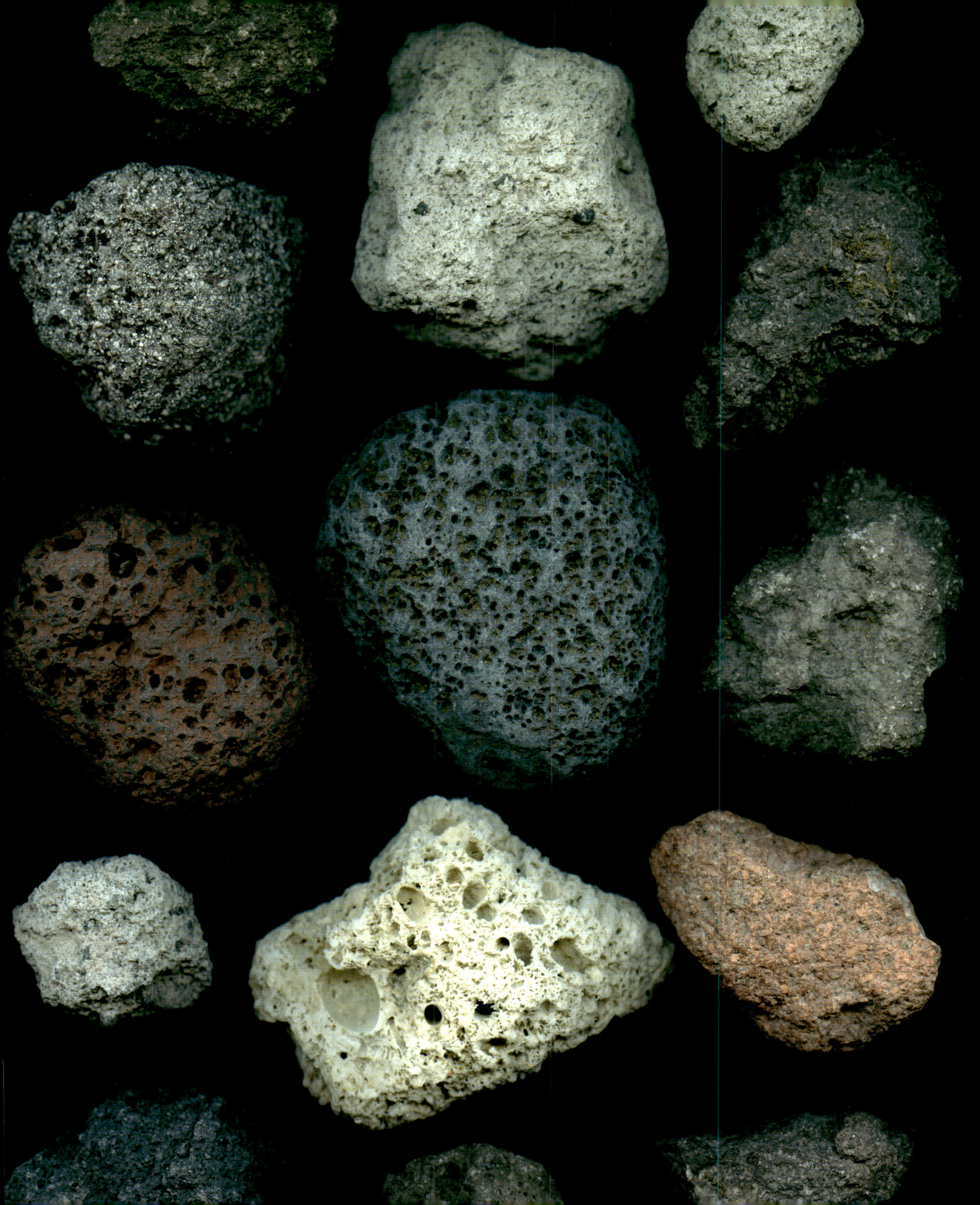

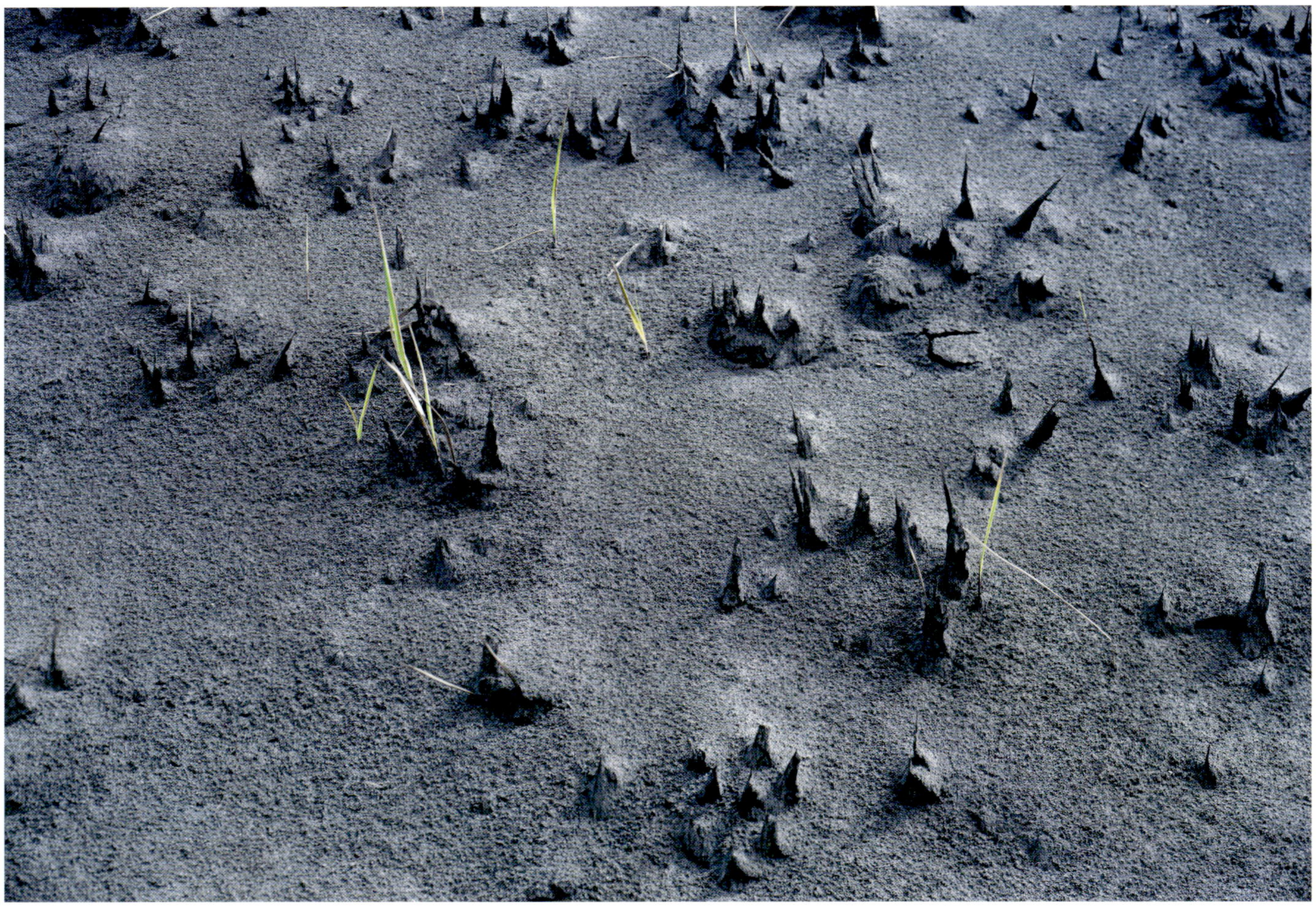

Reemerging Grass, Mt. Bromo *Volcano*, East Java, Indonesia, 2011, 18 ⅝ × 27 in.

Volcano Cycle

Volcano Cycle explores deep time with photographic images of volcanoes from Indonesia's Ring of Fire. These evoke earth forces, climate change, and human coevolution. Large-format negatives have been printed digitally on prepared aluminum plates with archival pigment inks through either the dye sublimation or direct pigment printing processes. The metal echoes the deep timbre of the eruptions, as well as the transmutation of materials, minerals, and metals that occur during a volcanic event.

After eons of fire and ice and then melt, our Eden emerged. Without the earliest volcanic activity, enough carbon dioxide would not have been released to produce global warming to melt the ice cover. Oxygen could then arise to eventually breathe us into being.

Mount Toba erupted on Sumatra seventy-four thousand years ago, creating an extreme volcanic winter that extinguished, if not all of us, then a large portion of life. It is possible that some of us, who were lucky enough to have made tools, were able to crawl out of the deep ash, to keep going. Thus Mount Toba is referred to by scientists as the "Weak Garden of Eden," because, at least at Mount Toba, life had to begin anew.

Volcanoes made it possible for life to exist on Earth.

They enable us to see the beauty of the great natural forces of the planet—of which we are part. They offer a window into both our own evolution and participation in global warming. I want people to see how alive the Earth is without it being terrifying. These pictures are the opening of a conversation.

All works in this section are dye sublimation or archival pigment on aluminum

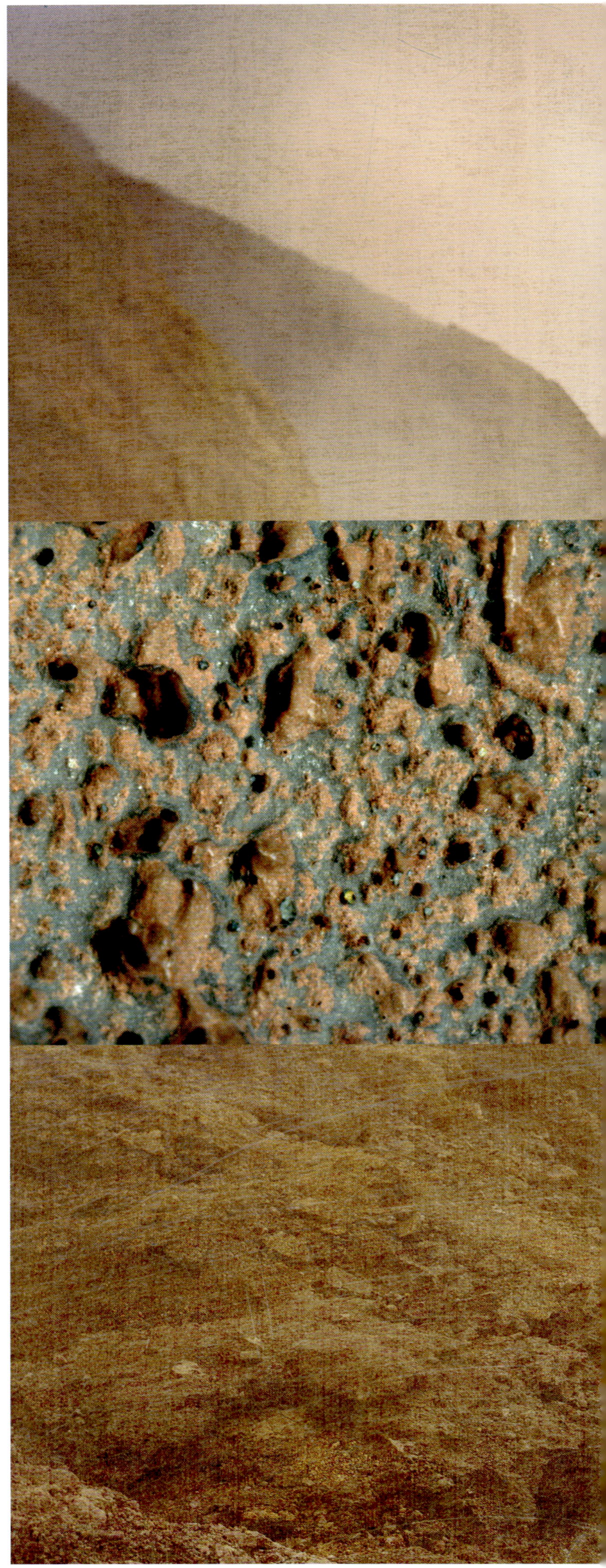

Ring of Fire, 9 plates, 2011–2013, 43 ½ × 55 ½ in.

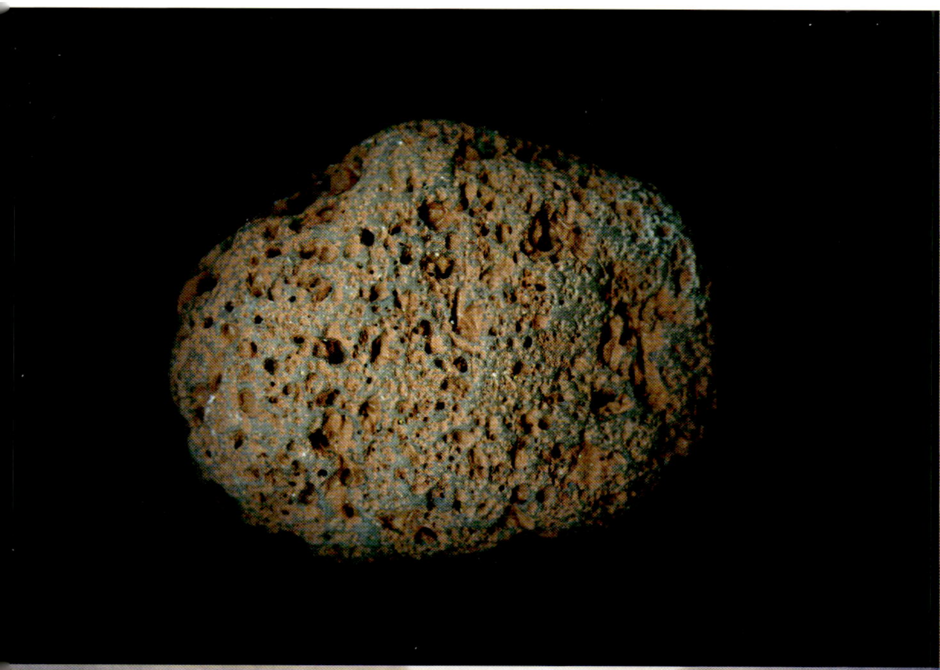

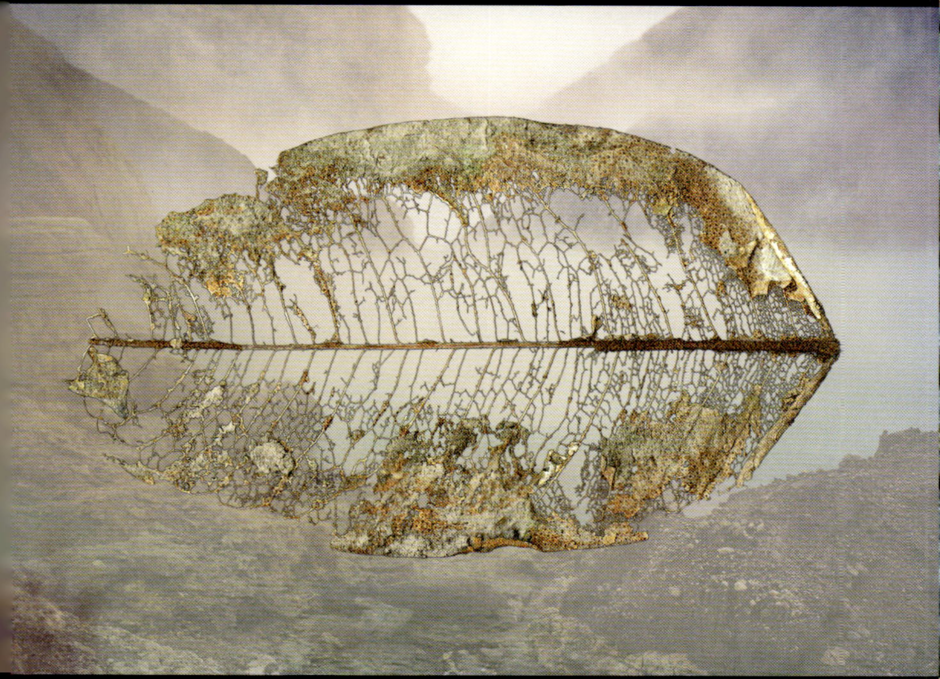
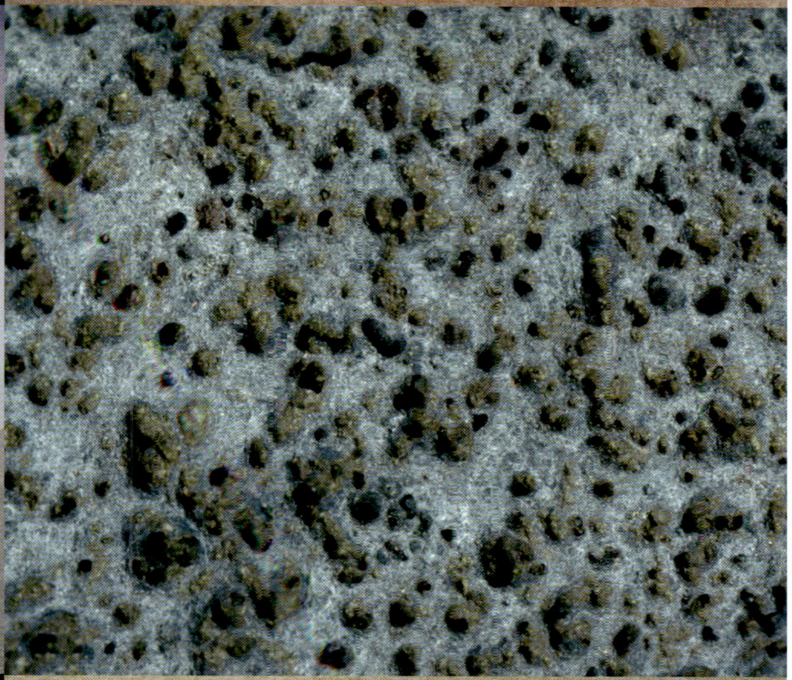
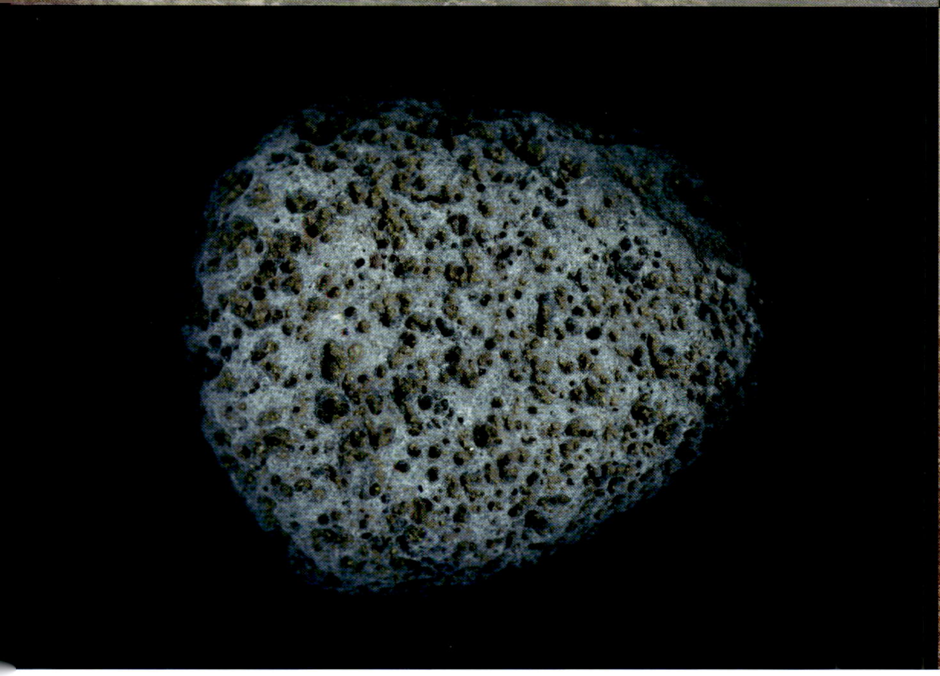
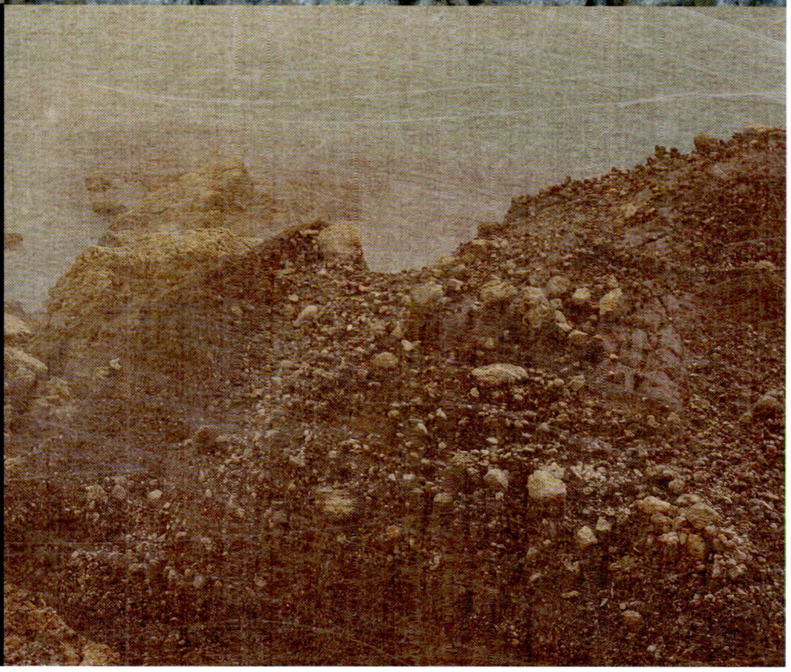

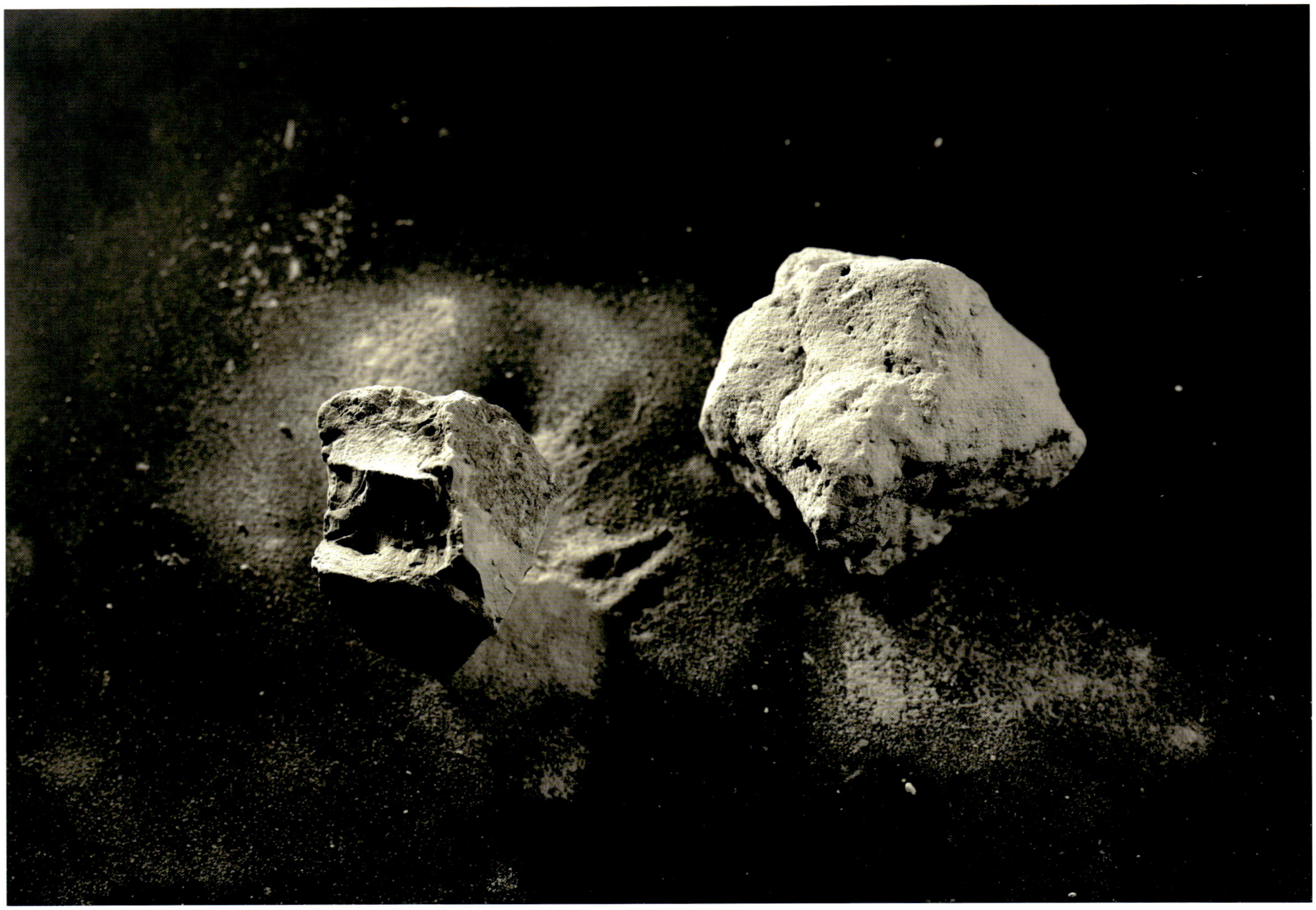

Mt. *Toba Volcanic Ash, 74,000 Years Old*, from Sumatra, found in Malaysia, 2010, 18 ⅝ × 27 in.

right: *Between Heaven and Earth*, Mt. Bromo *Volcano*, East Java, Indonesia, 2011, 43 ⅝ × 21 in.

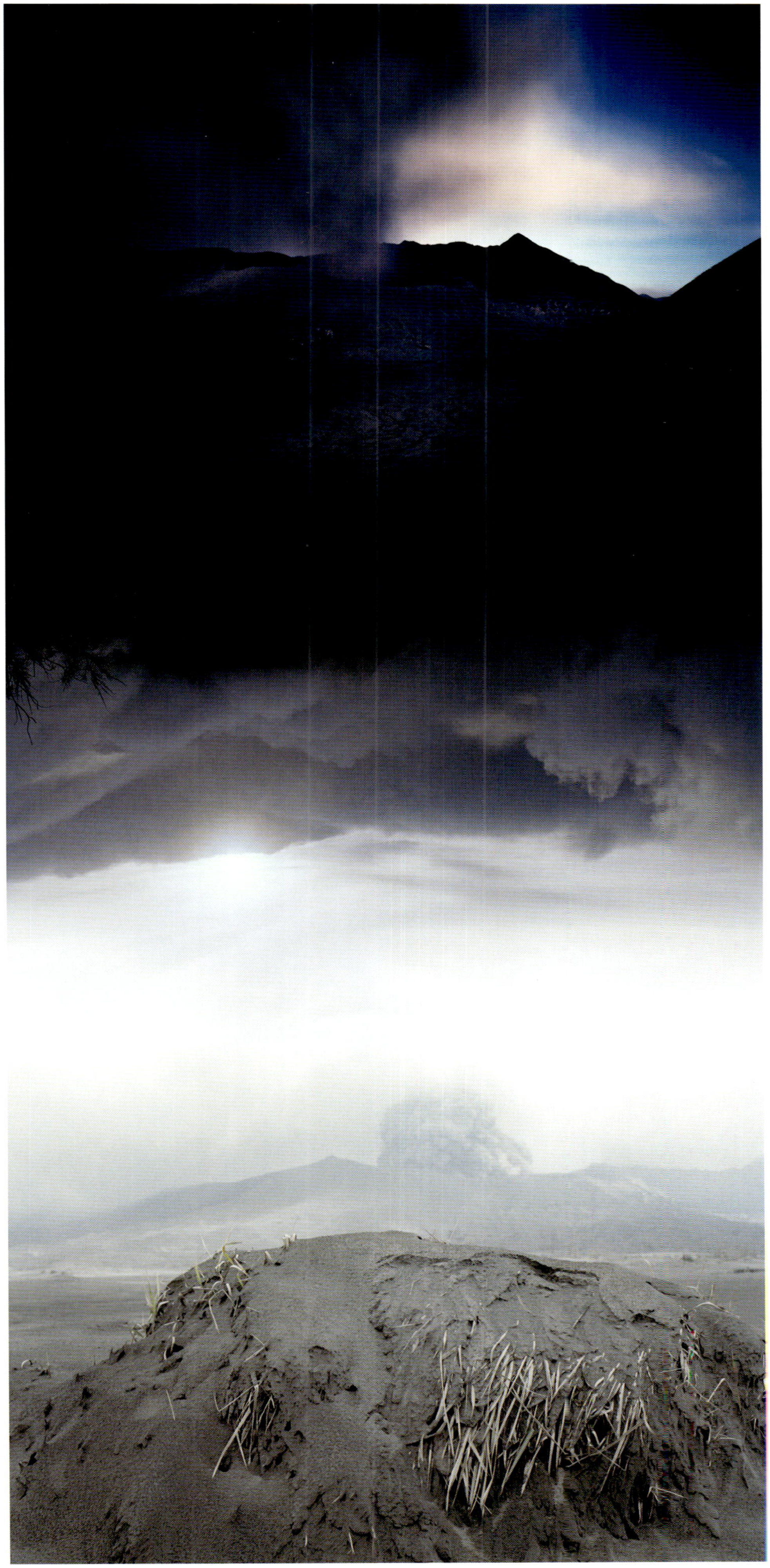

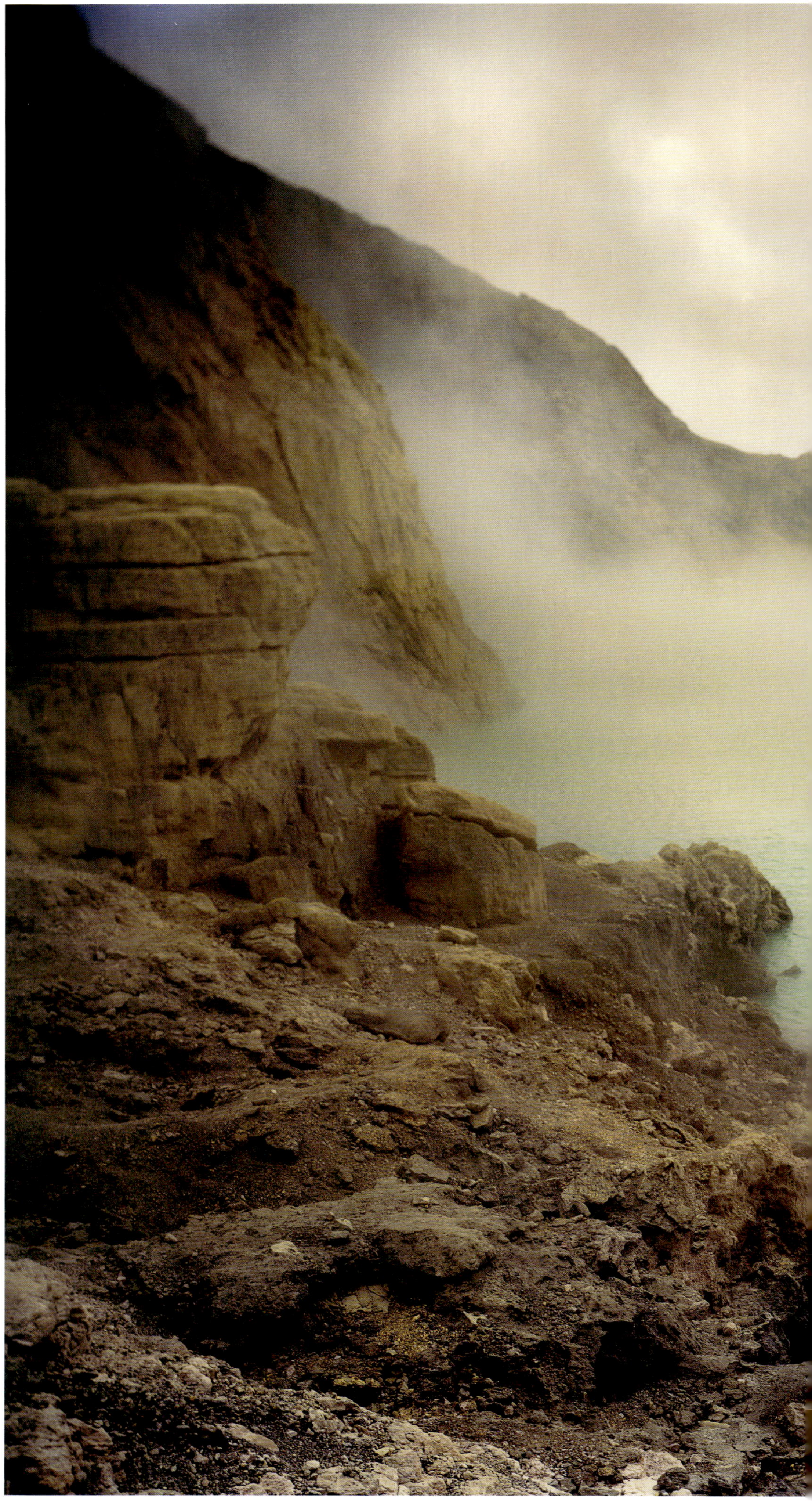

Acid Crater Lake, Kawah Ijen Volcano, East Java, Indonesia, 2011, 30 × 43 ¾ in.

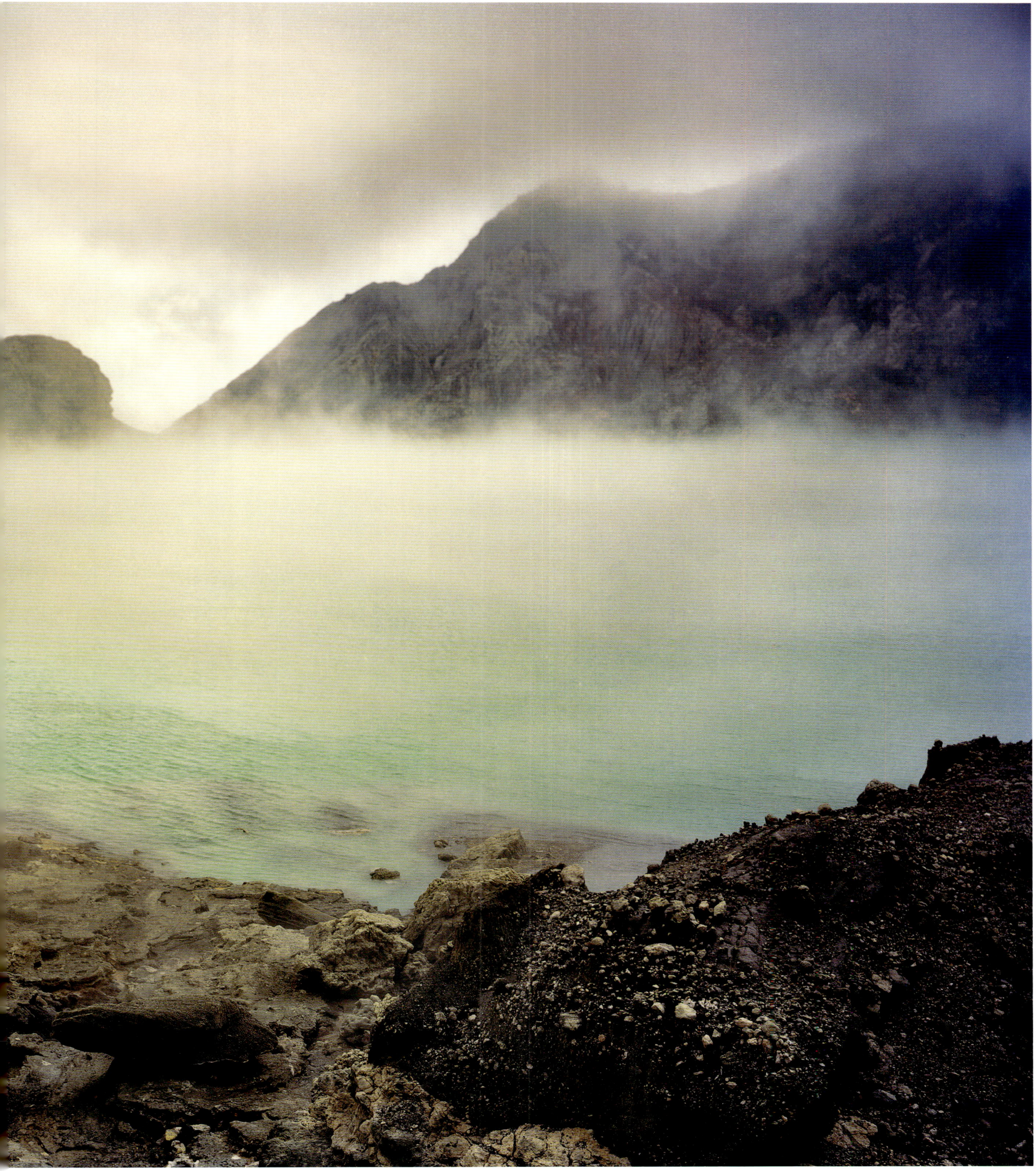

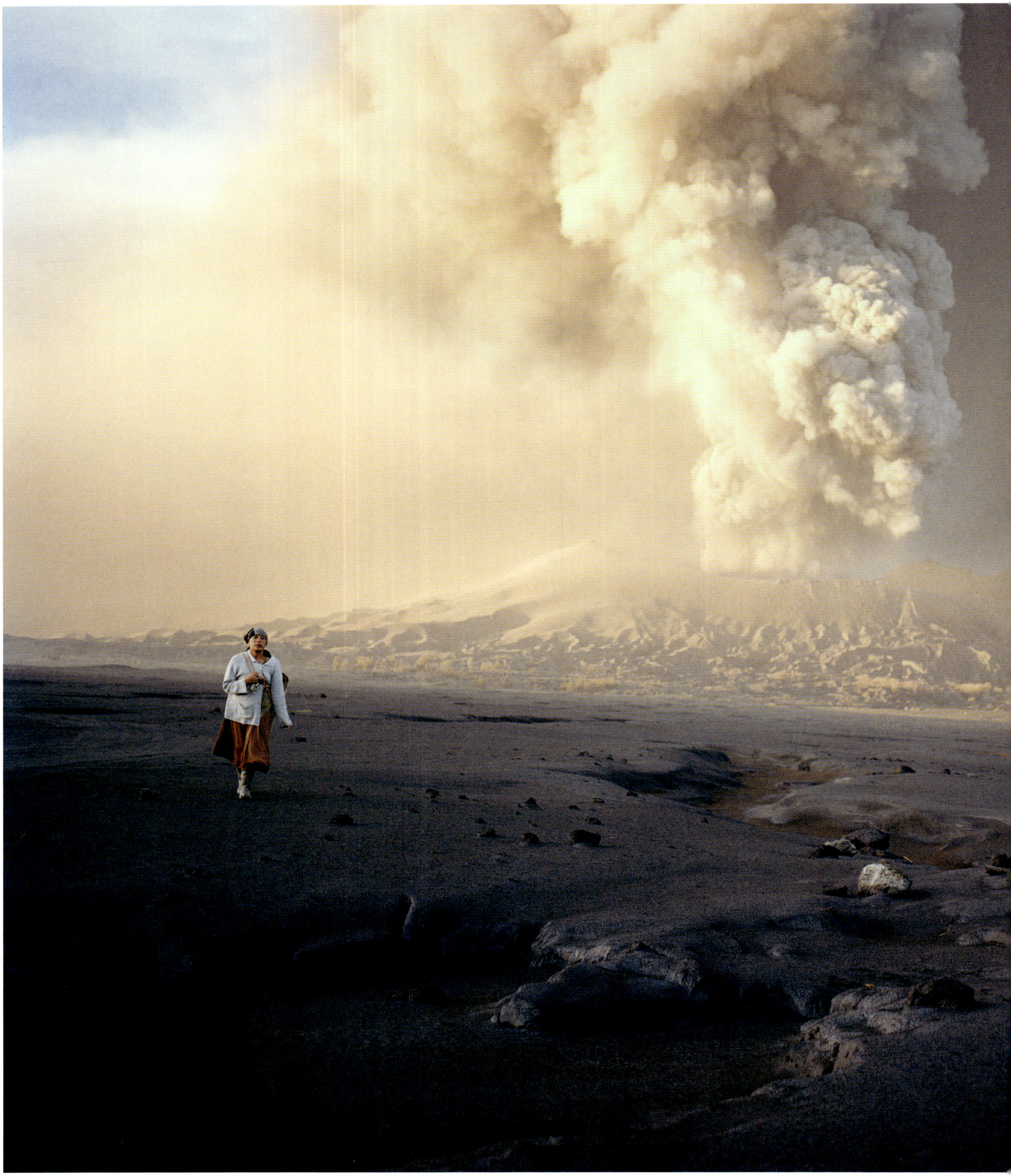

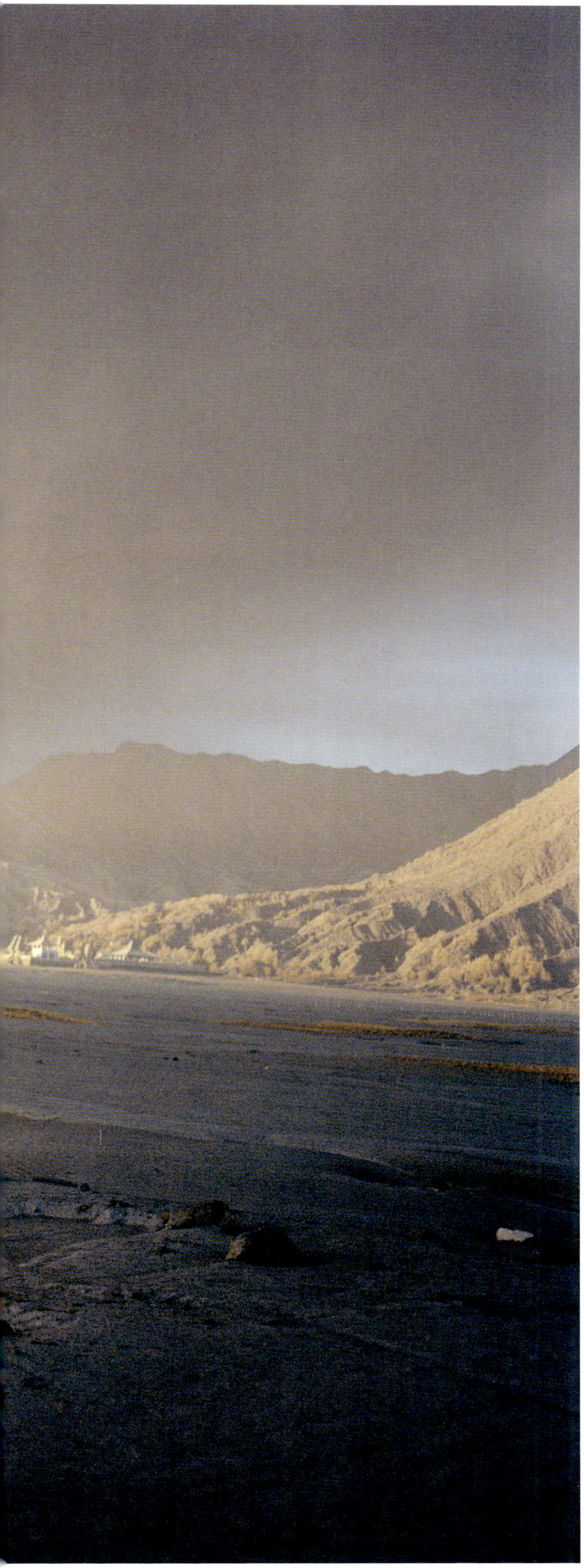

Lelani at Dawn, Mt. Bromo, 2011, 18 ¾ × 27 in.

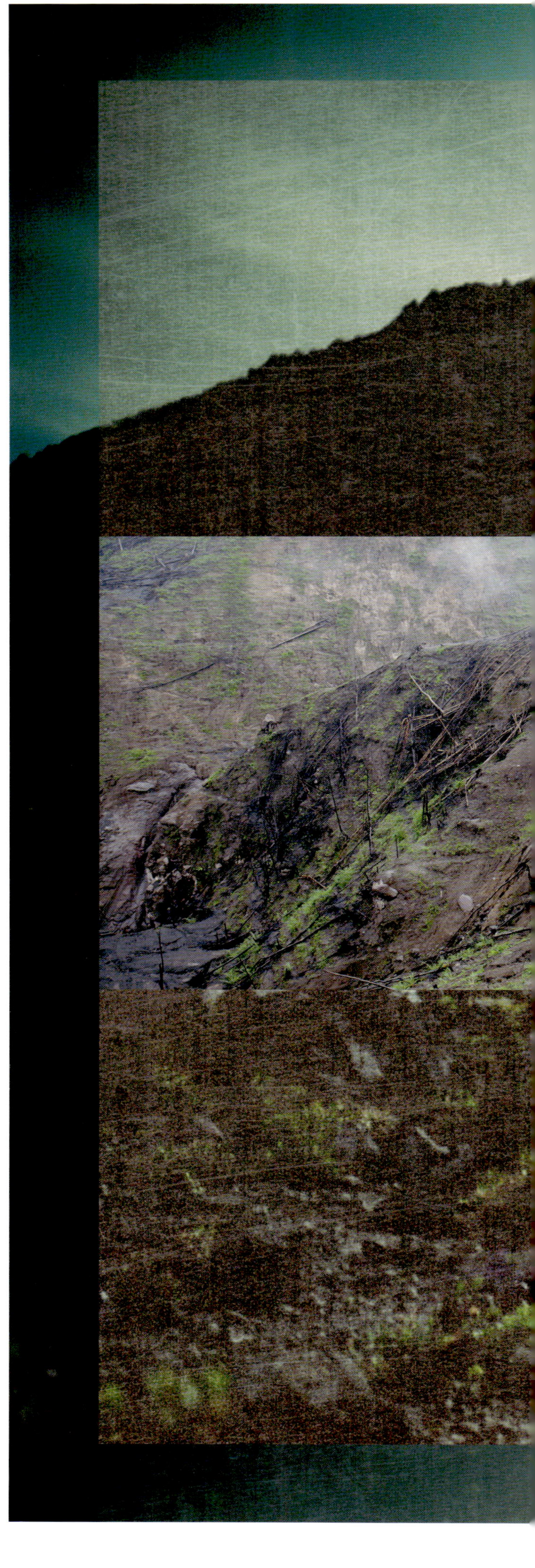

Pyroclastic Remains, 2012–2015, 43 ½ × 51 ¼ in.

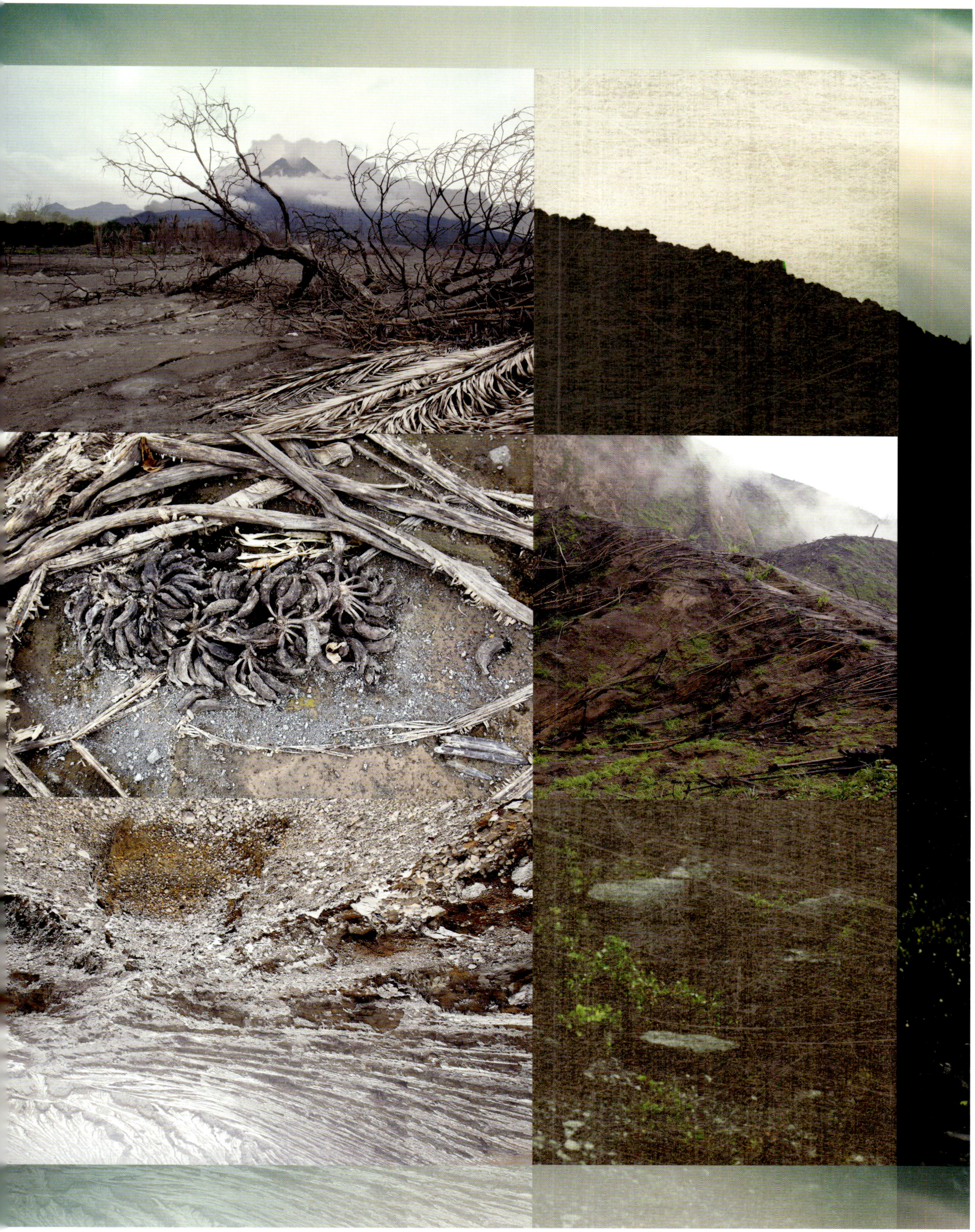

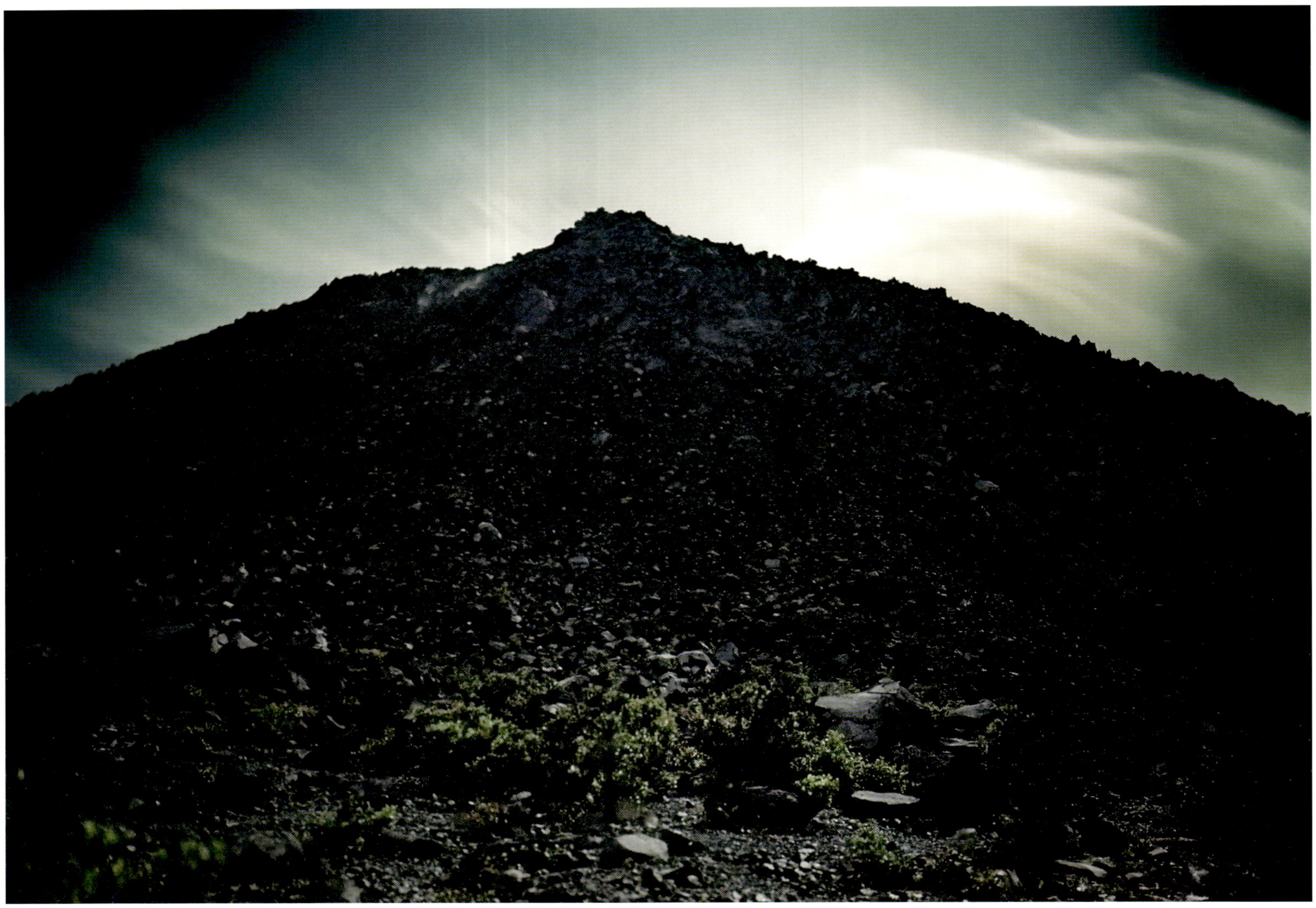

Mt. Merapi in Moonlight, the Calm before the Storm, 2010, 18 ⅝ × 27 in.

right: *Lelani, Mt. Bromo Volcano, East Java, Indonesia*, 2011, 27 ¾ × 18 ½ in.

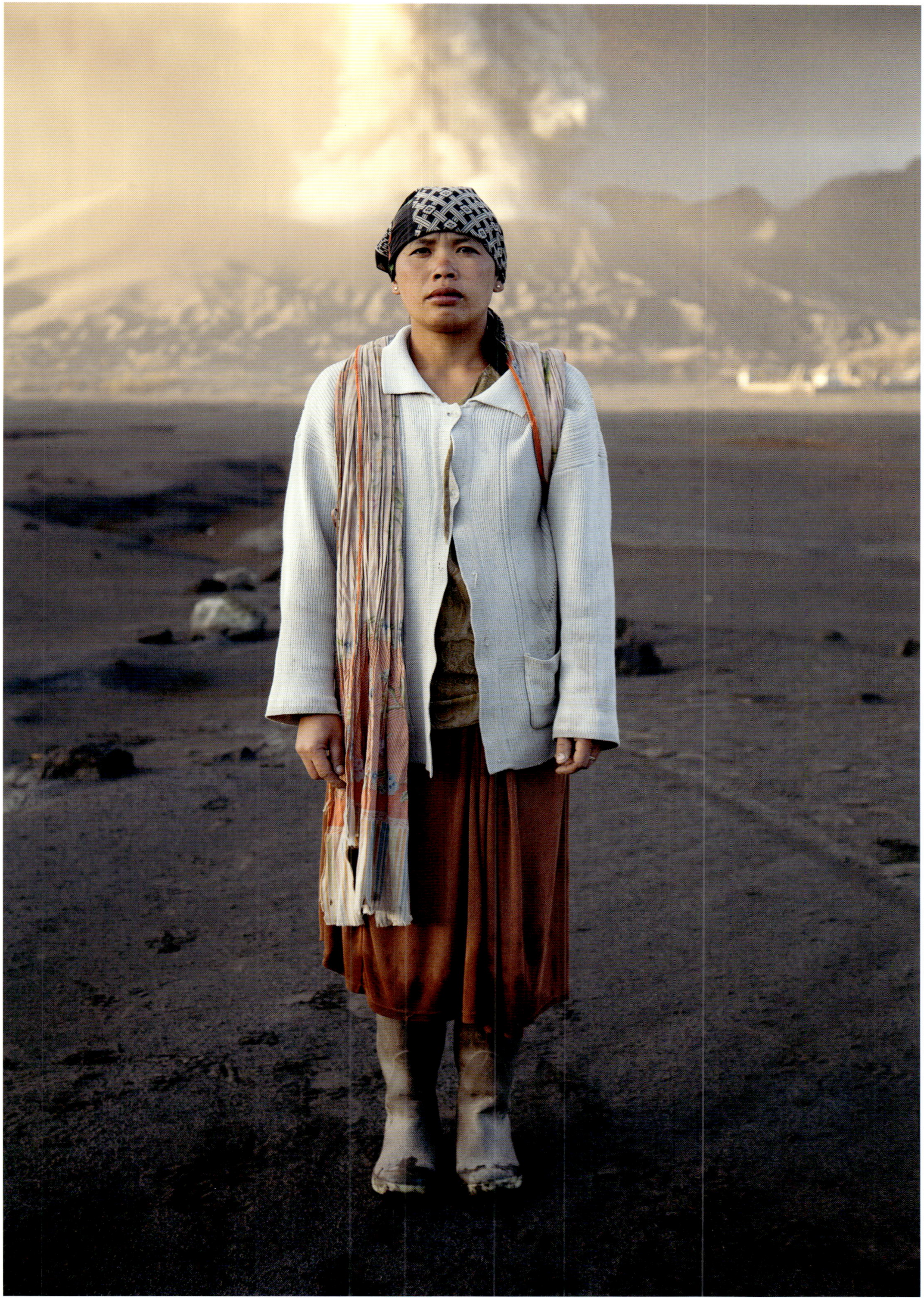

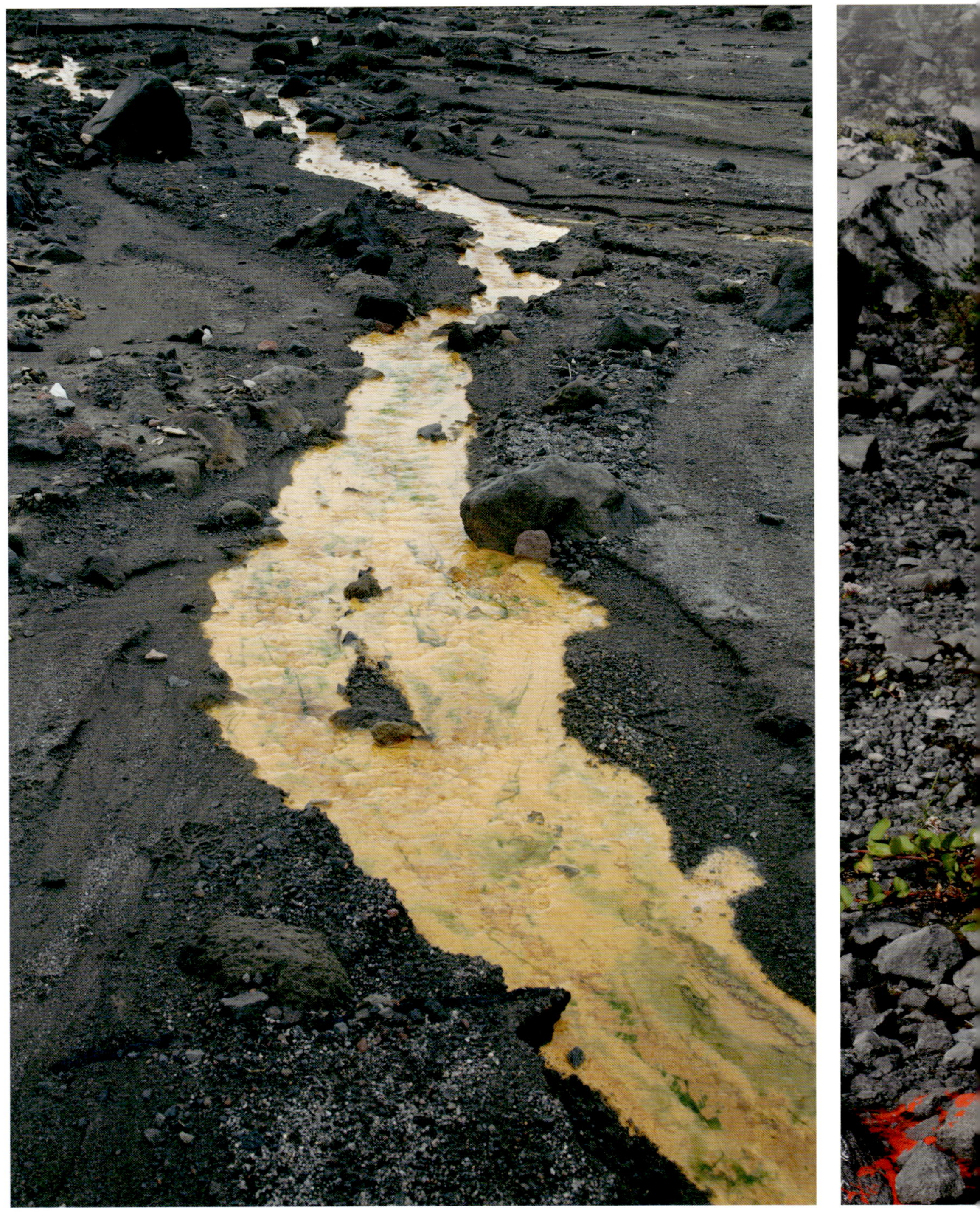

Pyroclastic Dance, 2010–2012, 28 × 57 in.

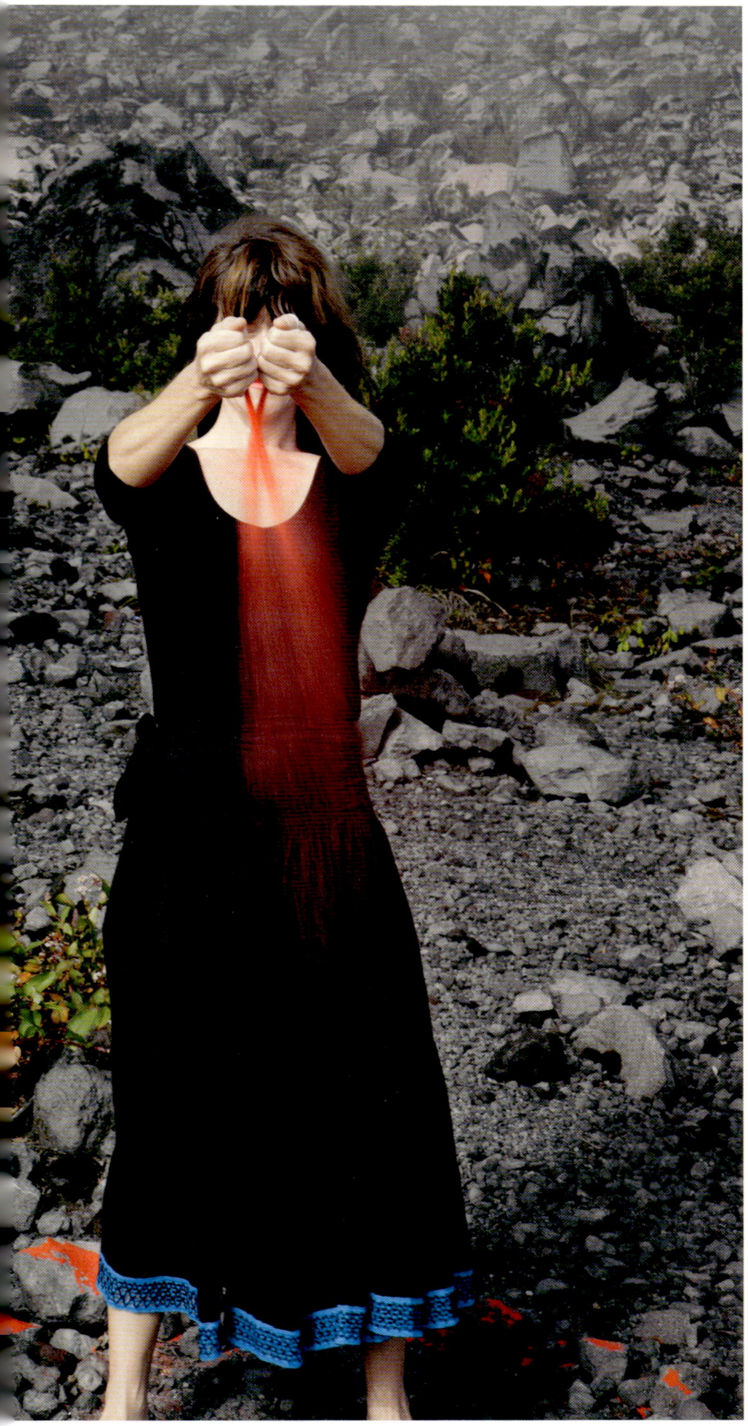
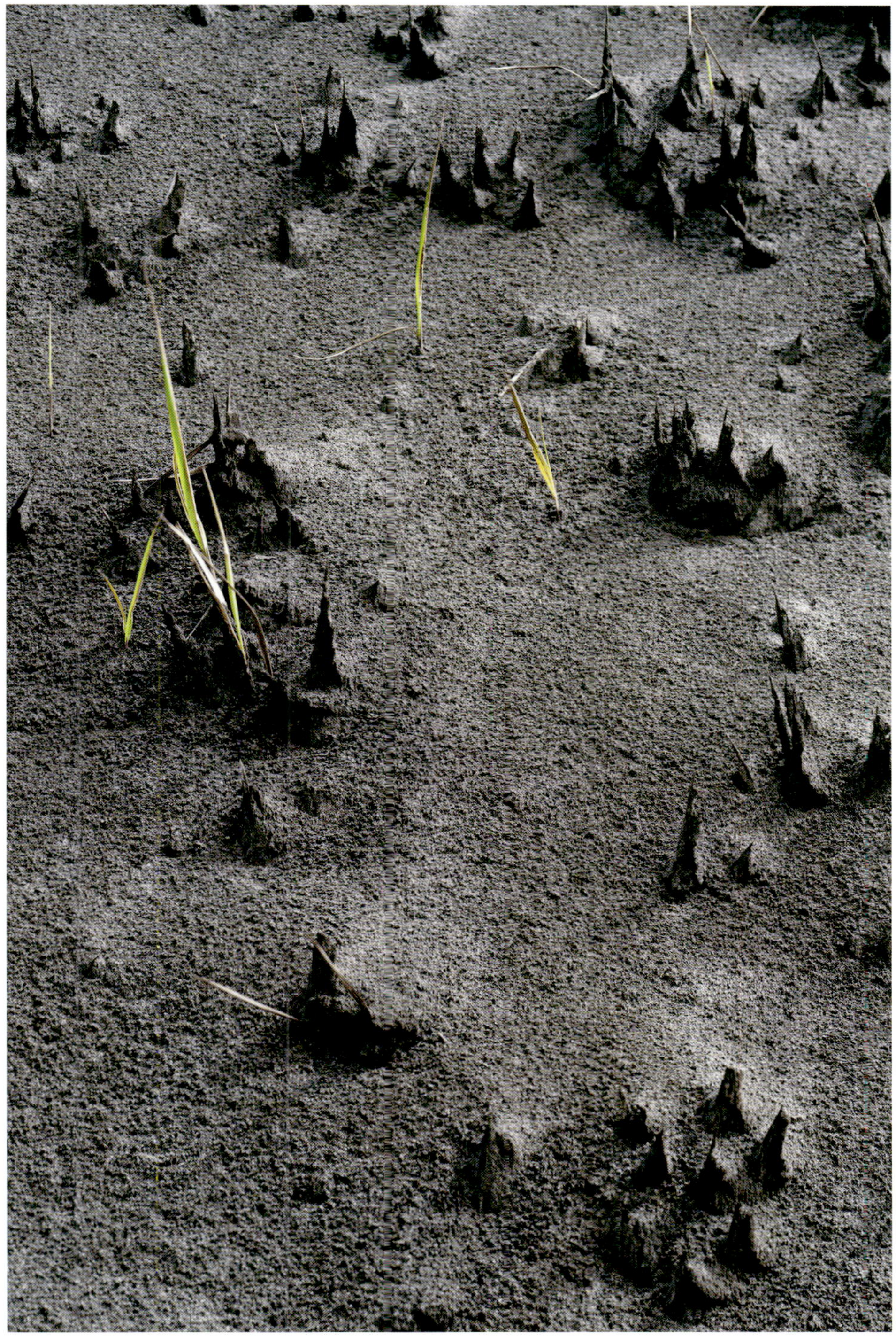

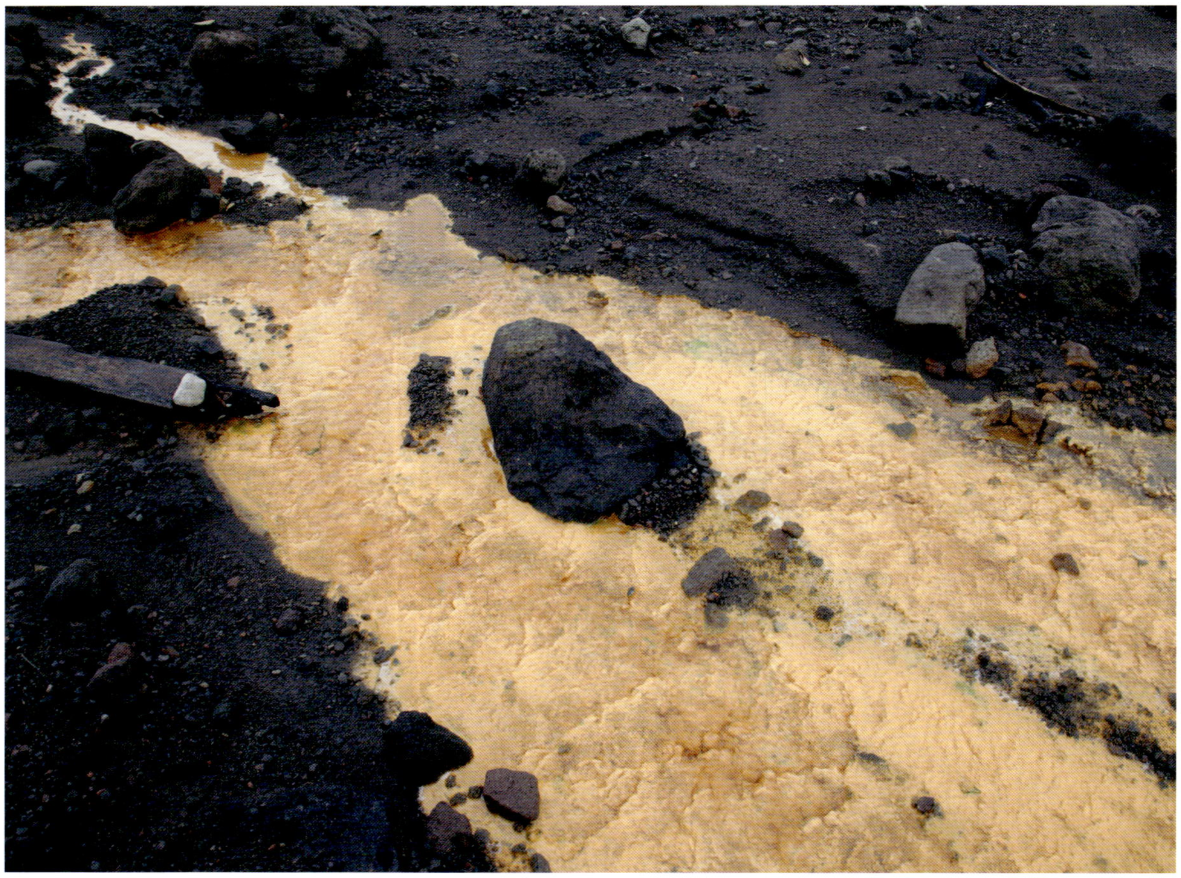

Iron Red River, after Mt. Merapi Eruption, East Java, Indonesia, 2011

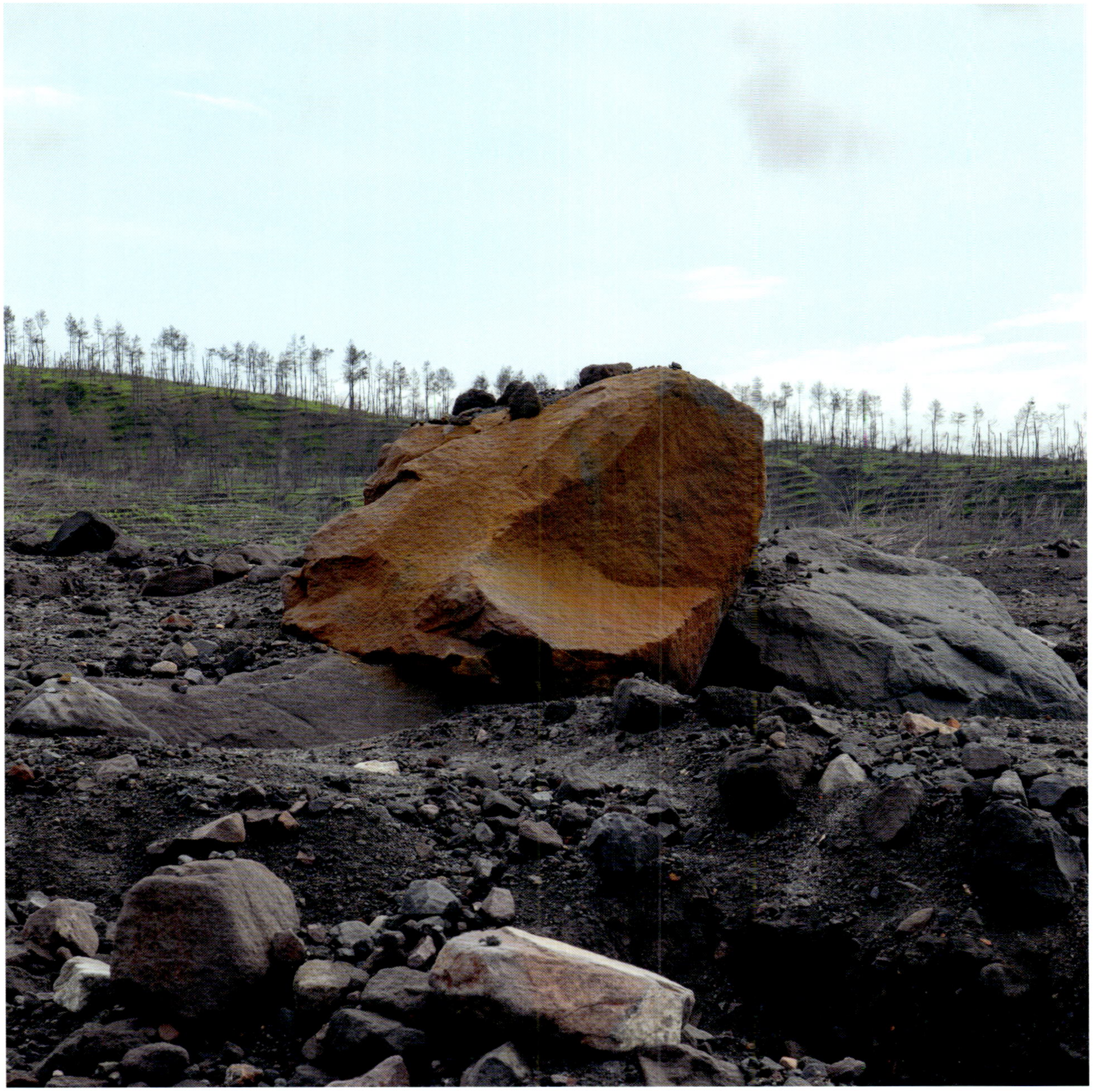

Red Rock, after Mt. Merapi Eruption, East Java, Indonesia, 2011

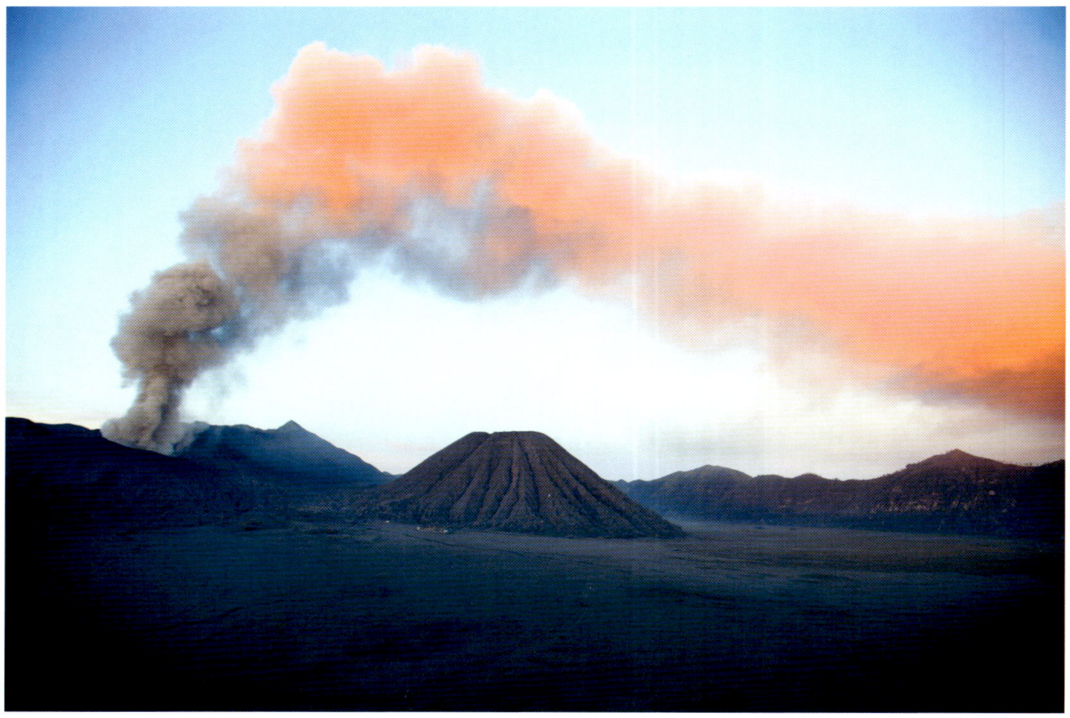

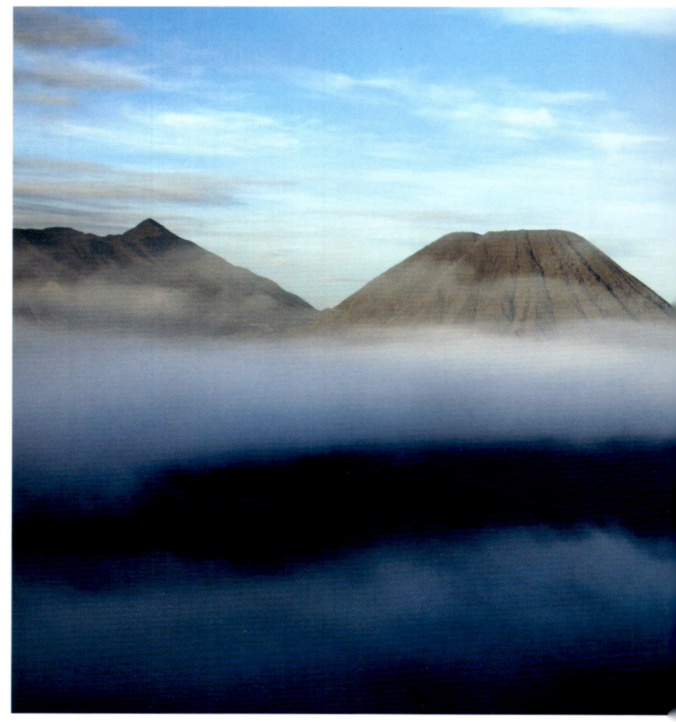

left to right: *Pink Trail*, *Mt. Bromo Volcano*, East Java, Indonesia, 2011, 18 ⅝ × 27 in.

White Trails, *Mt. Bromo Volcano*, East Java, Indonesia, 2011, 18 ⅝ × 27 in.

Mt. Bromo Sunrise, East Java, Indonesia, 2011, 18 ½ × 27 in.

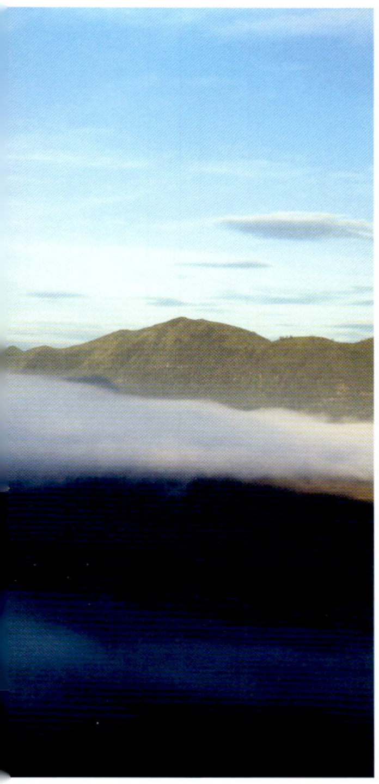

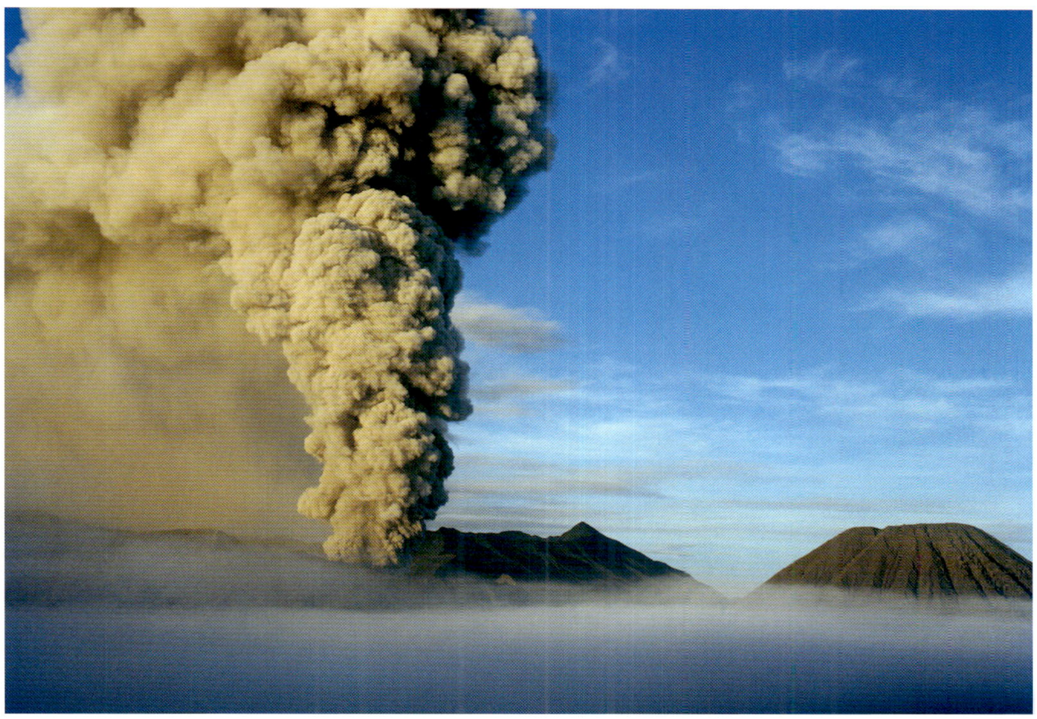

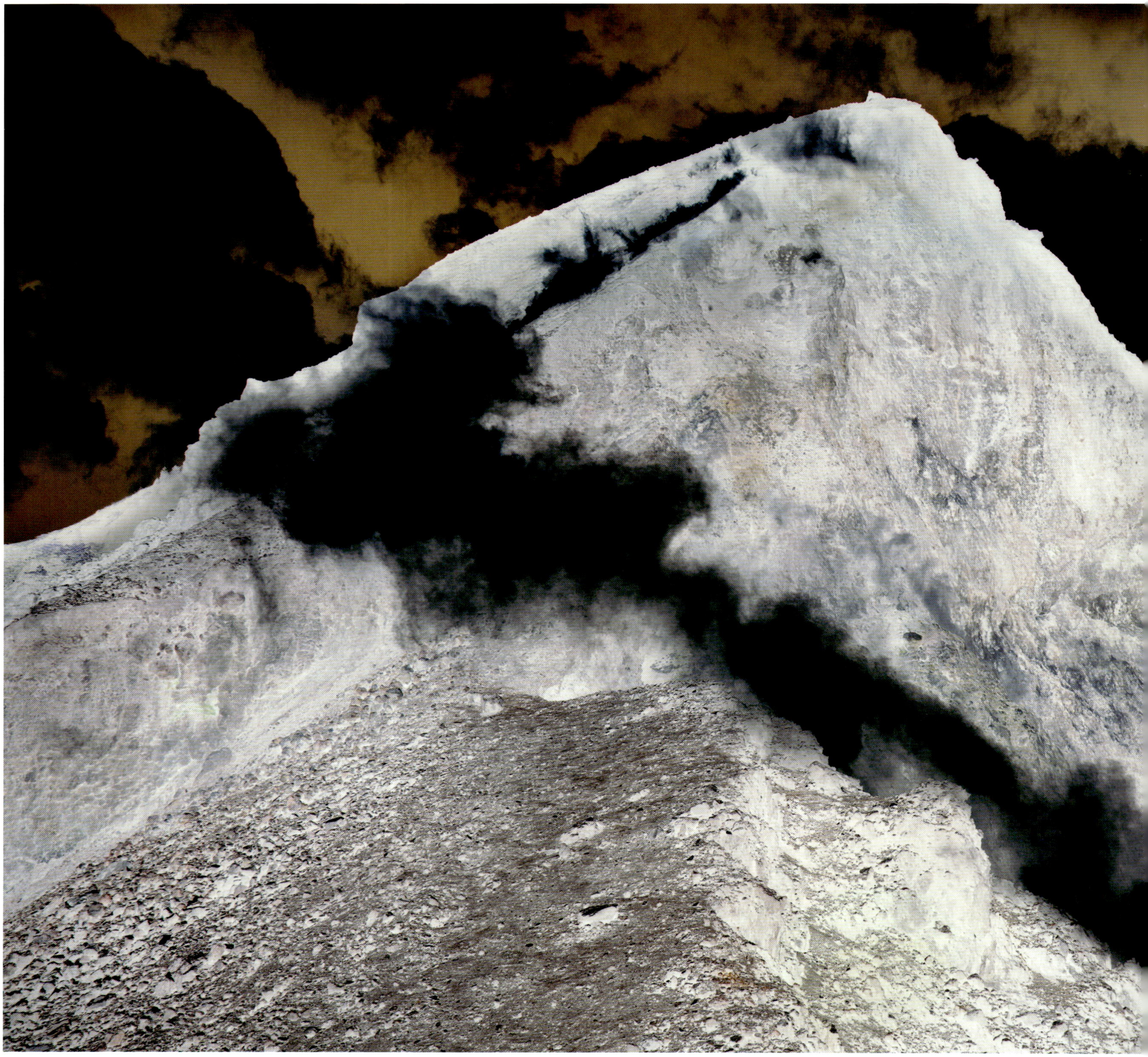

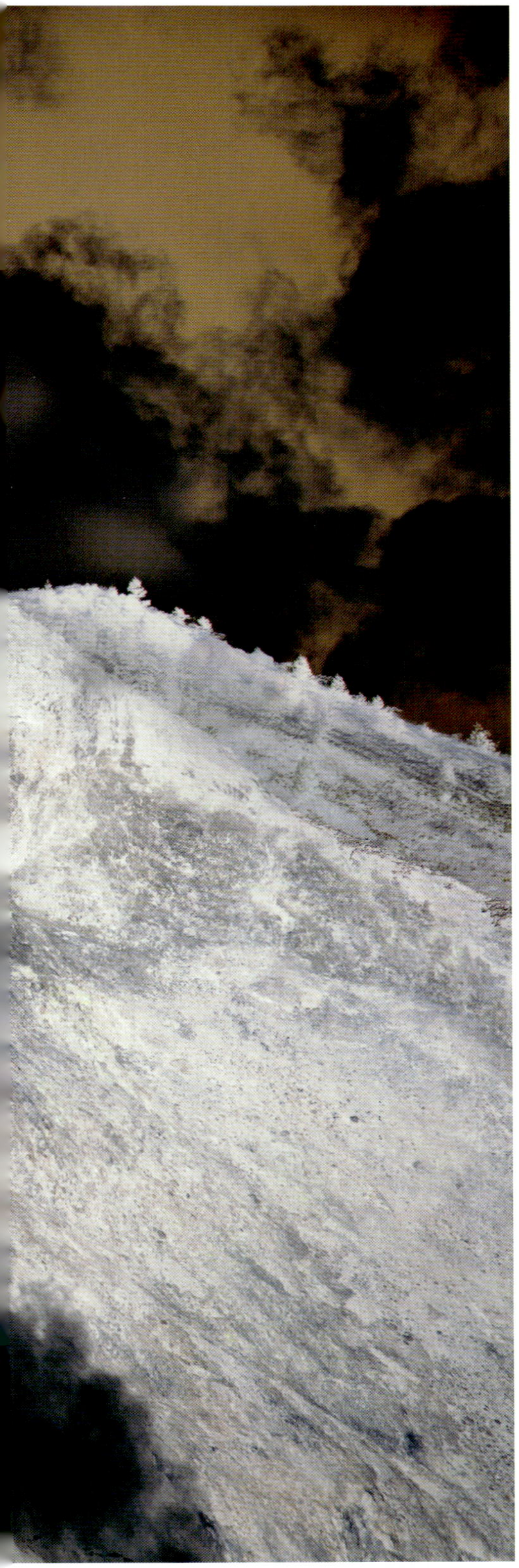

Mt. Batur, Bali, Indonesia, 2014, 30 × 43 ¾ in.

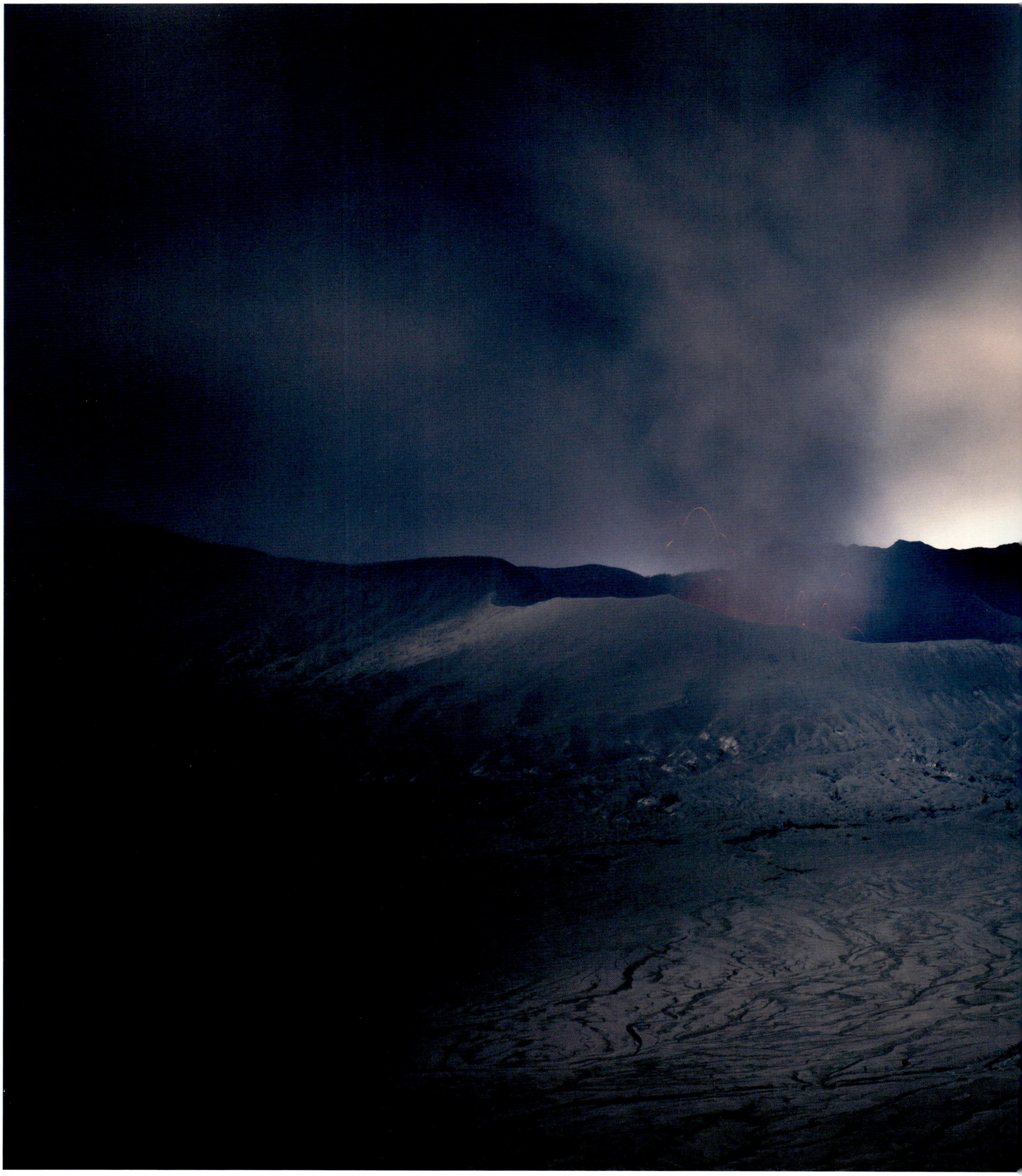

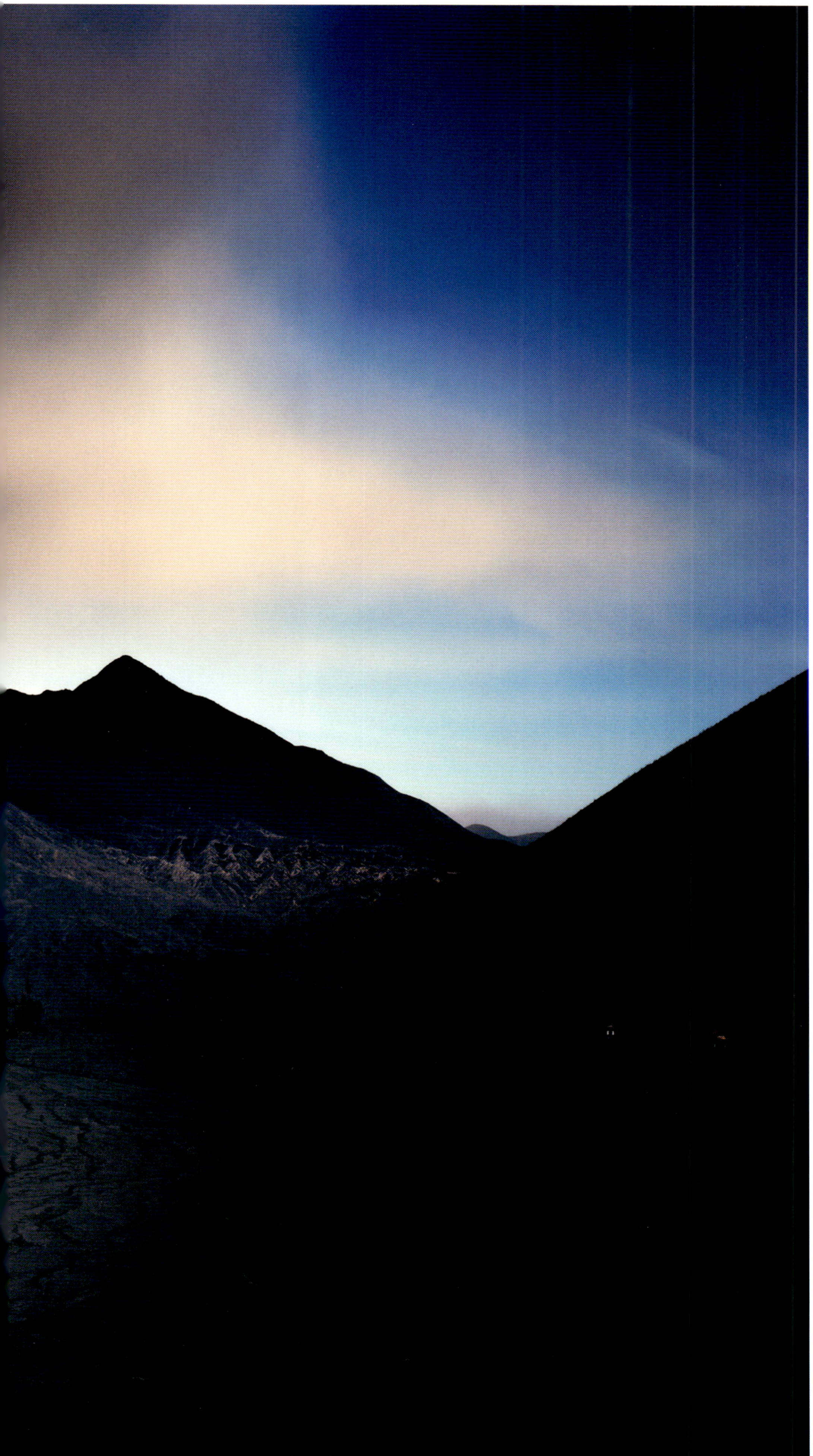

Mt. Bromo at Night, East Java, Indonesia, 2011, 18 ½ × 27 in.

following spread: *Mt. Bromo from above, Encircled*, Mt. Bromo Volcano, East Java, Indonesia, 2011, 30 × 43 ¾ in.

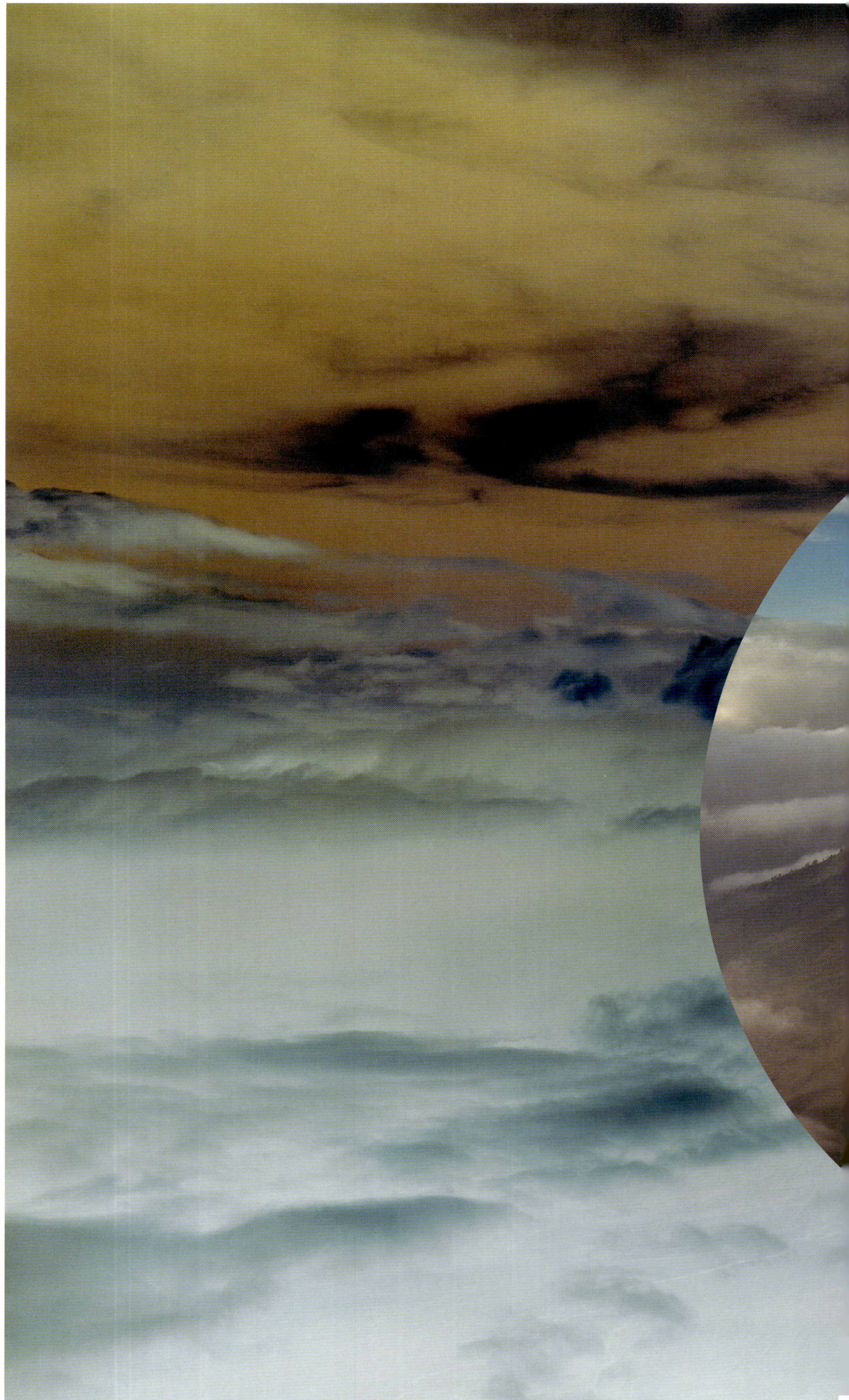

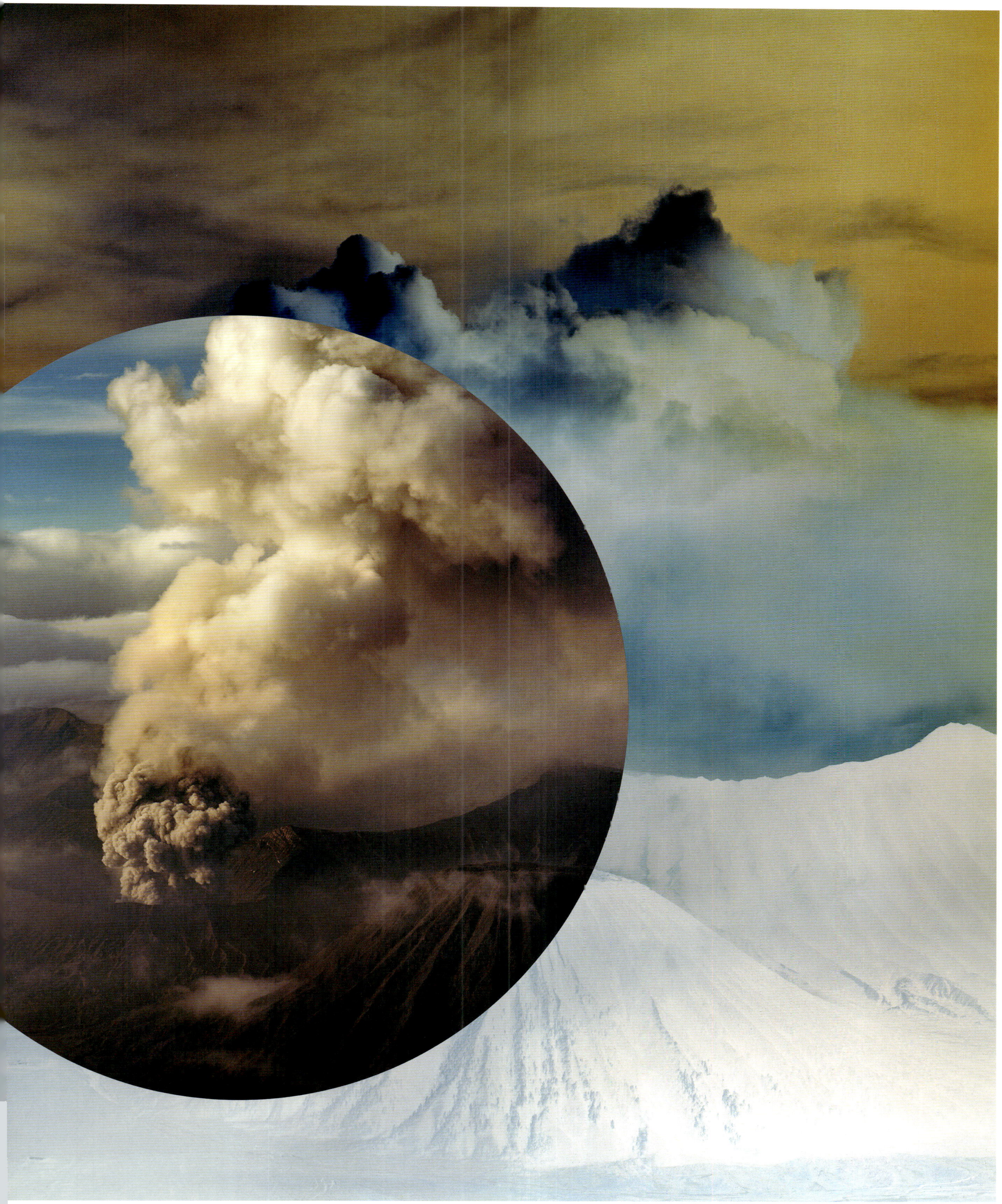

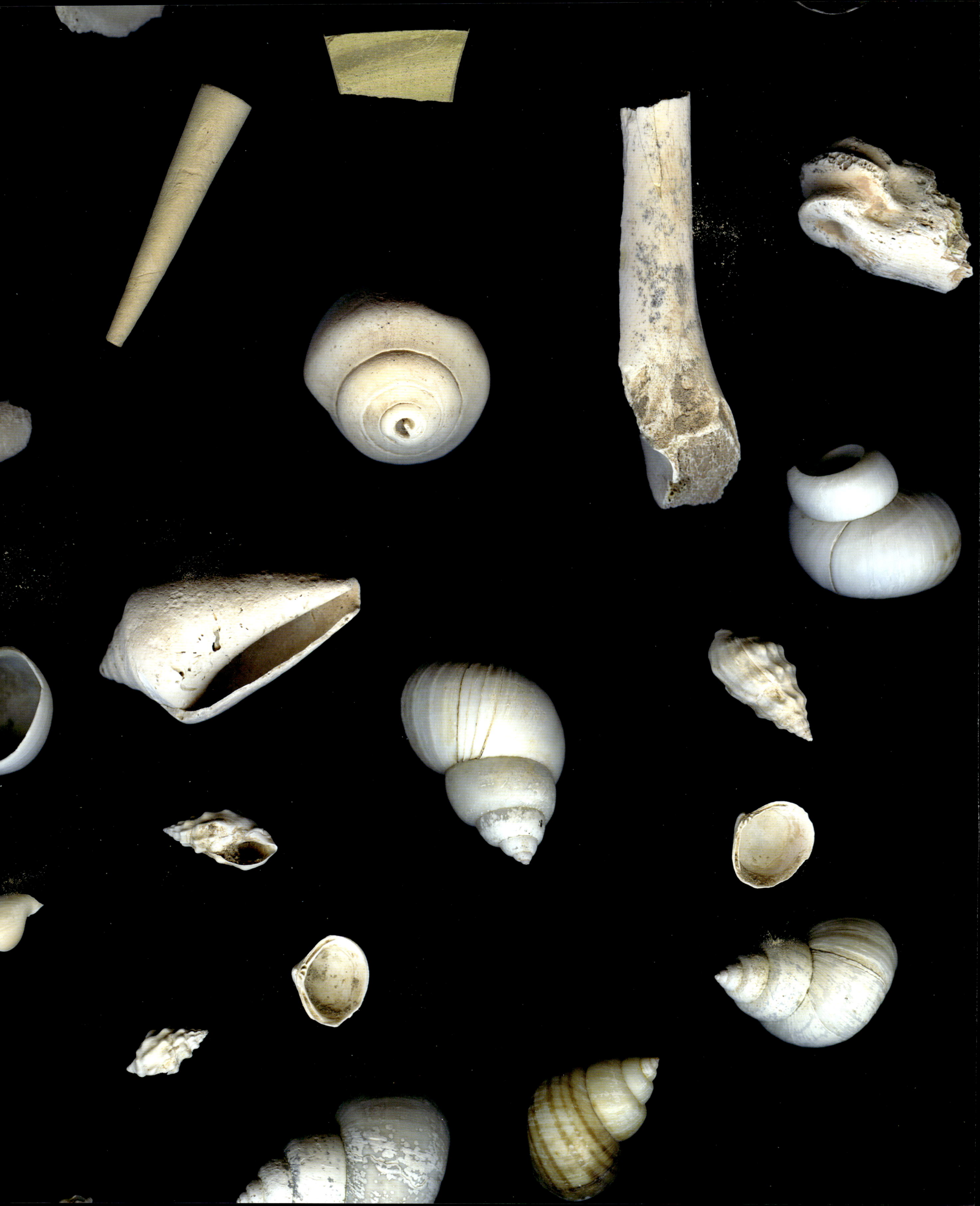

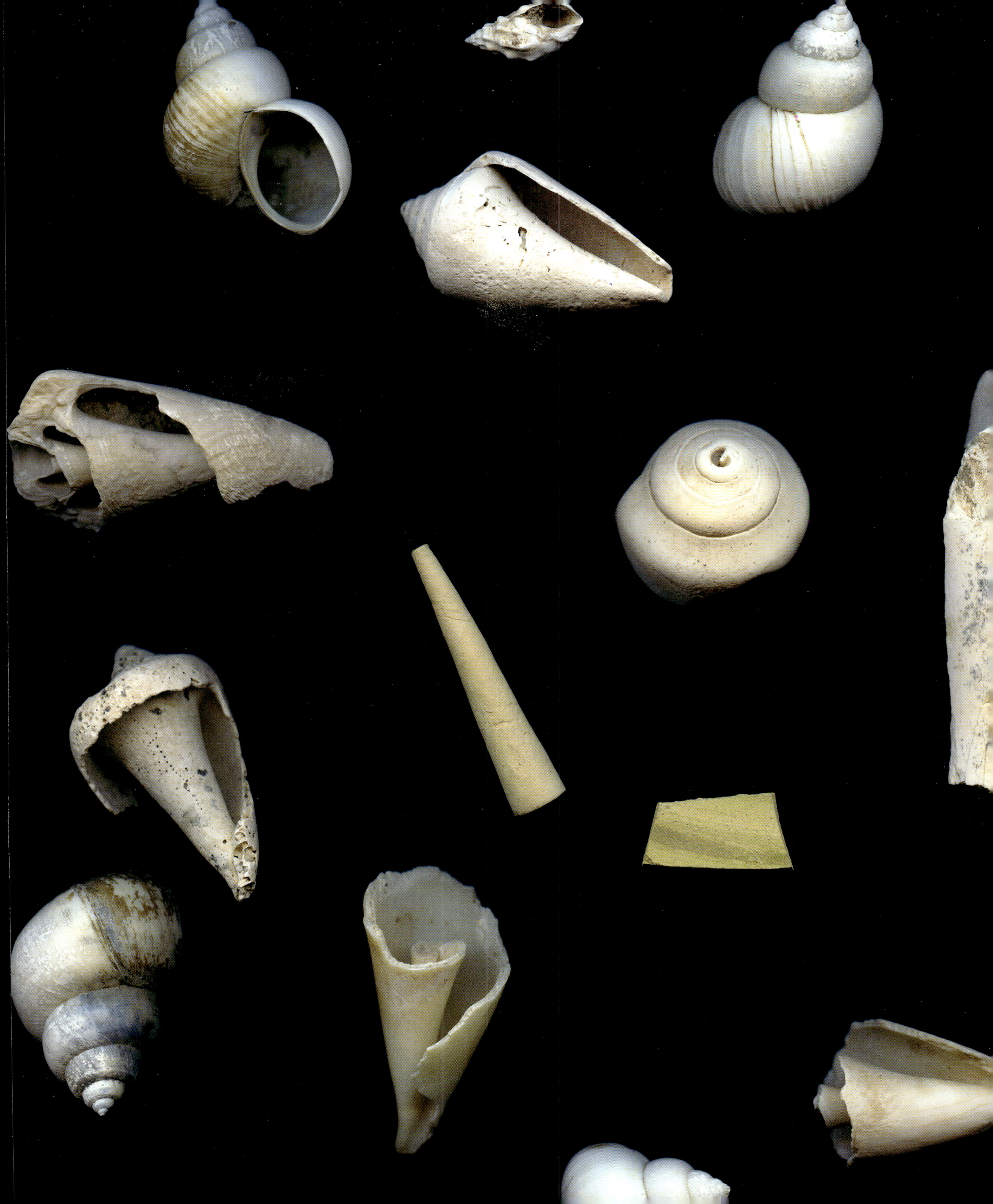

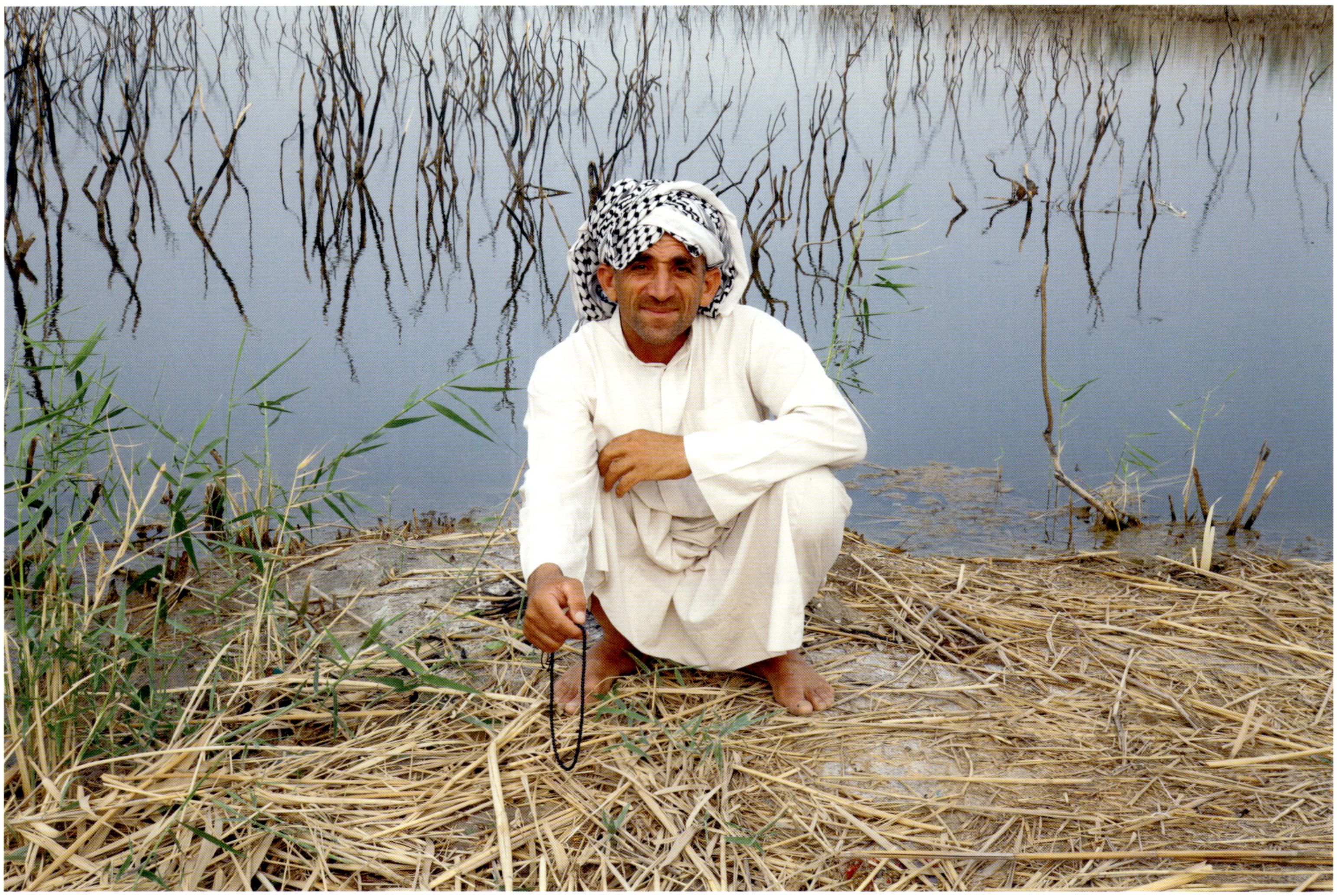

Ehmad on His New Mini Floating Island, Hammar Marsh, Southern Iraq, 2011

Eden in Iraq

In 2010, in the midst of creating both *Photosynthesis* and *Volcano Cycle* for *Eden Turned on Its Side*, I learned about the valiant efforts of the Iraqi Environmental NGO, Nature Iraq, to restore the vast drained Mesopotamian marshes of southern Iraq. The Tigris and Euphrates Rivers cross on the eastern edge of the marshes, so these wetlands are thought to be in extreme proximity to the original historic site of the Garden of Eden.

The marshes are home to the Marsh Arabs, who lived an unchanged way of life here for over seven thousand years, until the early 1990s. Then, when Saddam Hussein drained and burned the wetlands to punish Shi'ite rebels hiding there, they were murdered or driven out. With Sadam's demise in 2003, people started returning, with simple hand tools, to break through the earthen canal and let the waters back into the desiccated land. Azzam Alwash, an engineer from Laguna Beach, California, immediately returned to the nature preserve of his childhood. He created Nature Iraq with the mandate to help restore and preserve Iraq's endangered waters, flora, and fauna.

Despite progress, the situation is far from perfect—with people now building on land instead of on waterways, the Euphrates has become dangerously contaminated with human sewage. I initiated the *Eden in Iraq* project in 2011 to create a demonstration wastewater garden that will restore health and grow beauty in the devastated marshes of southern Iraq.

These images are made to resonate, both conceptually and in a material sense, with the other two parts of *Eden Turned on Side*—*Photosynthesis* and *Volcano Cycle*—and to remain companions to the *Wastewater Garden Project*. Our garden design is based on Mesopotamian wedding-blanket patterns. The images presented here are configured as weavings in the sense that different themes are being woven together: nature, culture, environmental and cultural history, and more. Here linen, tapestry, and metal are employed to give a sense of the fabric of life in the marhes.

All works in this section are printed on 100 percent rag paper, linen fabric, or aluminum, prepared specifically for digital prints with vegetable pigments as well as tapestries with the image woven directly into the material.

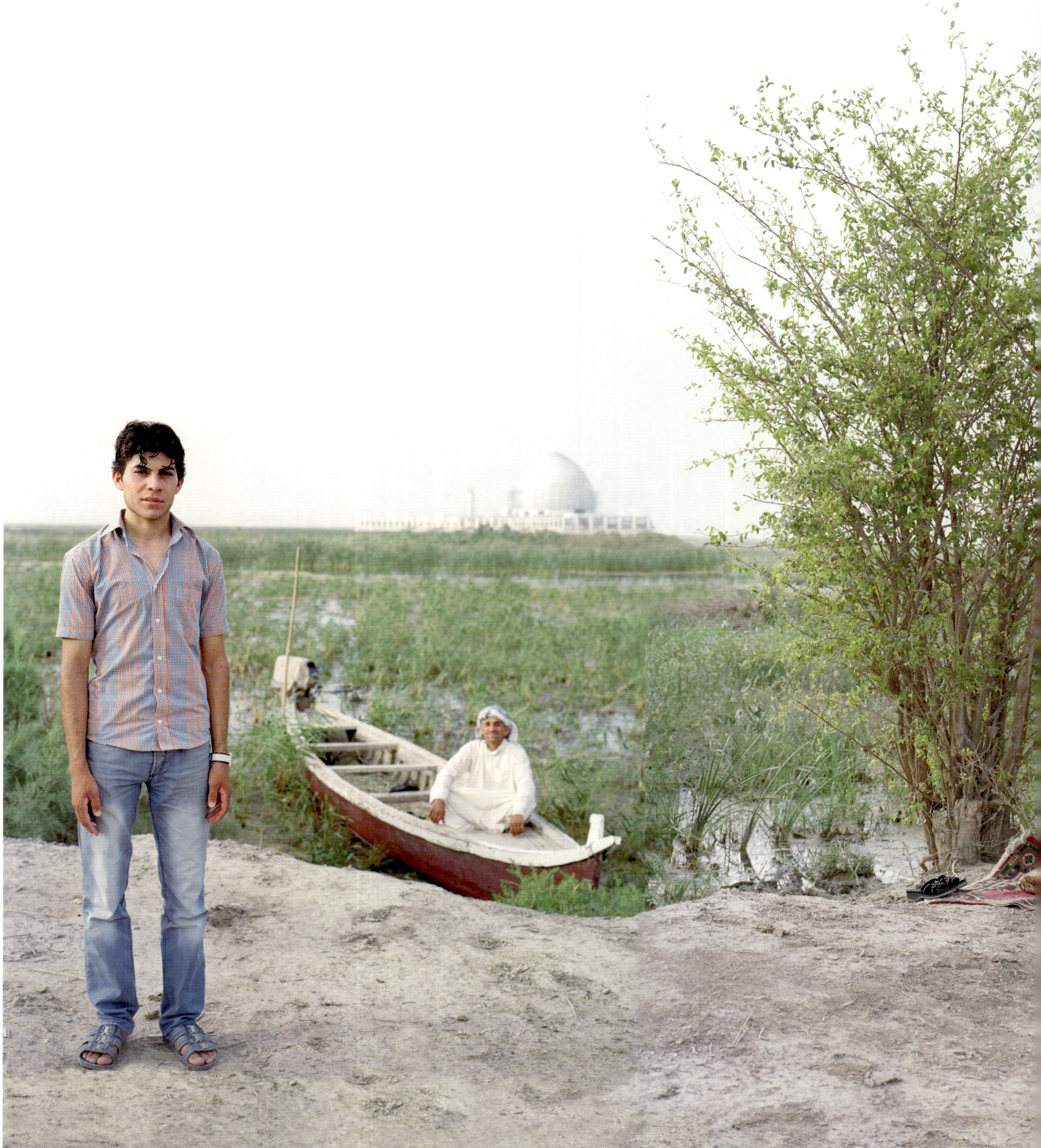

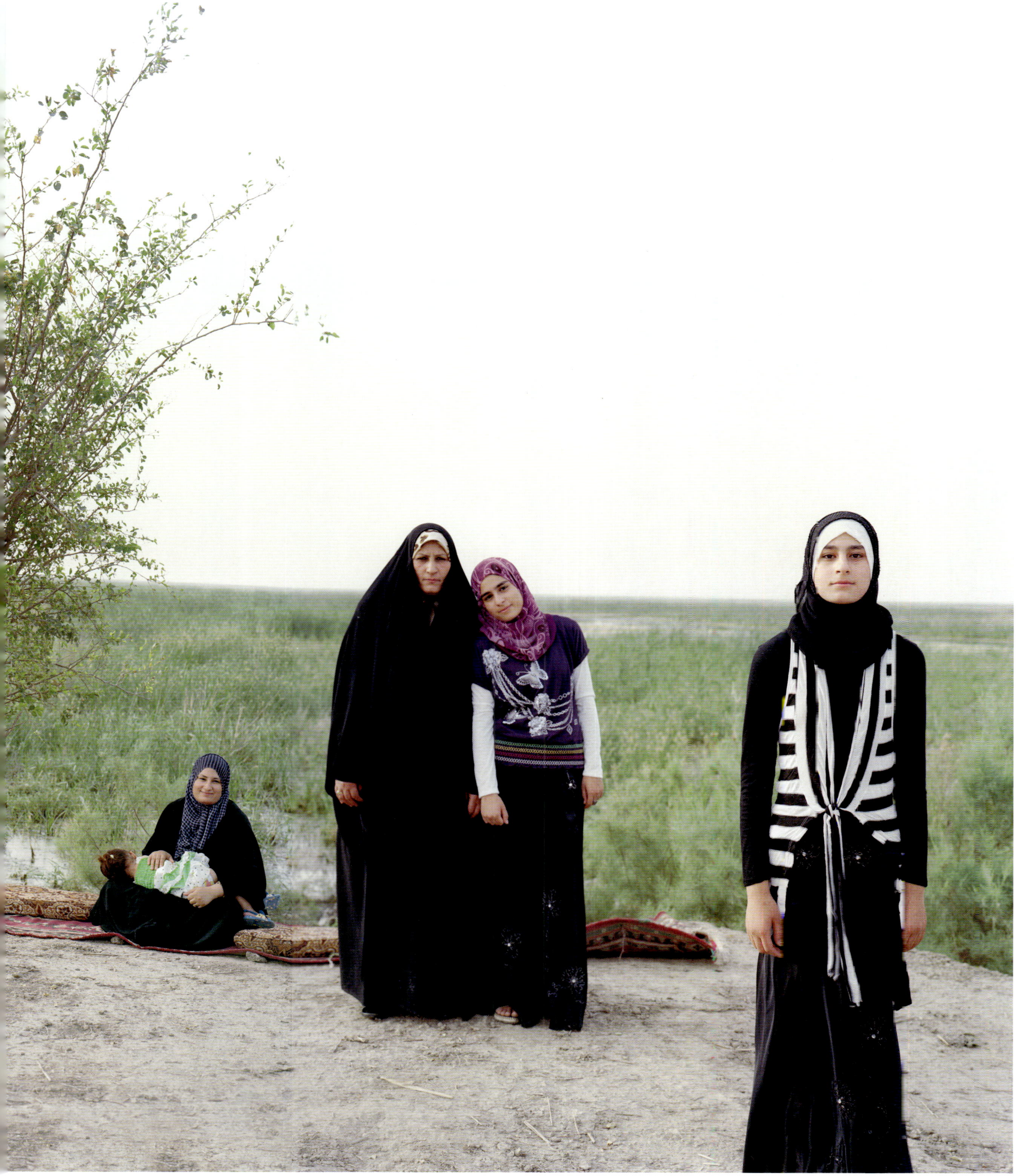

Adam and Eve in the Iraq Marshes, near the Possible Historic Site of the Garden of Eden, 2011–2012, 34 × 59 ¾ in., linen (41 × 66 in. with border)

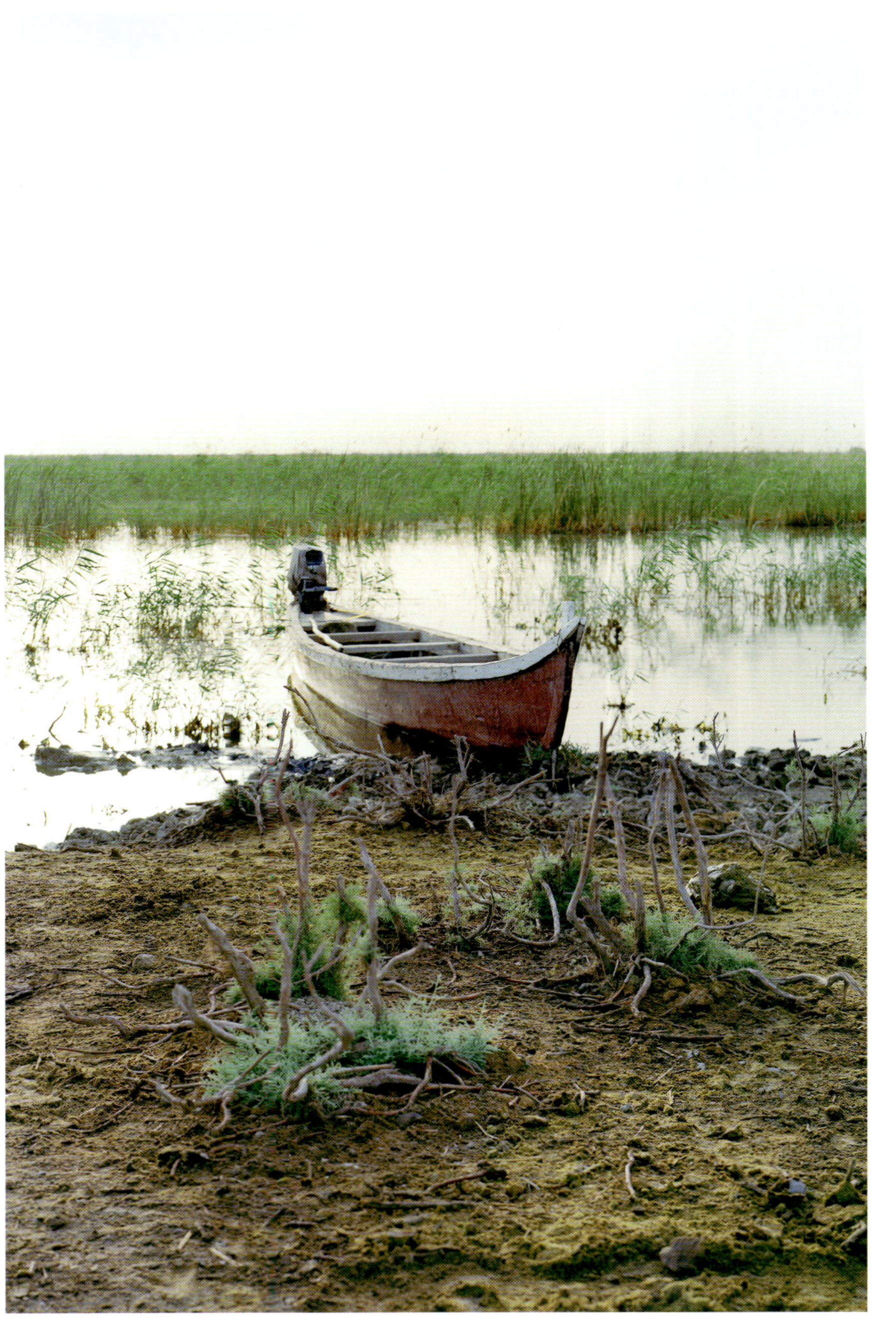

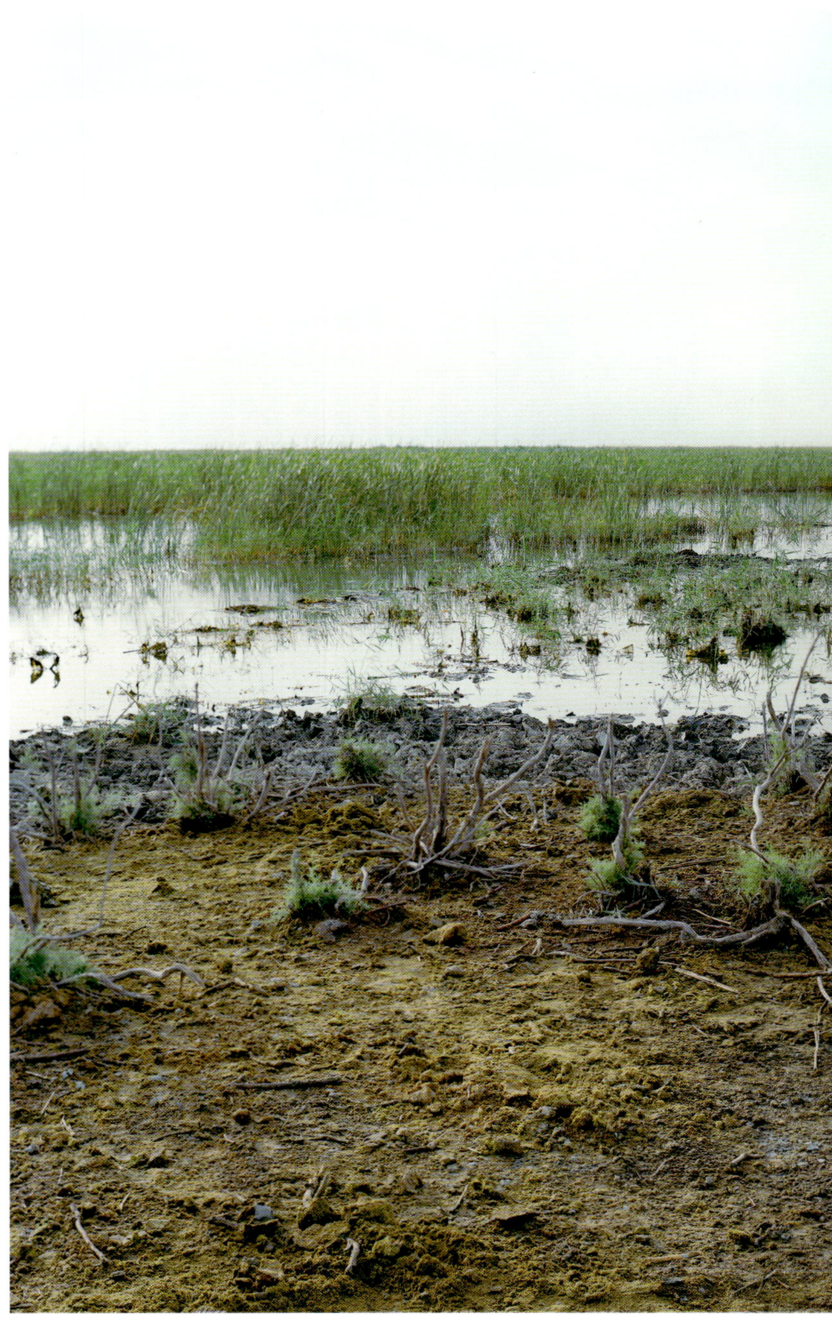

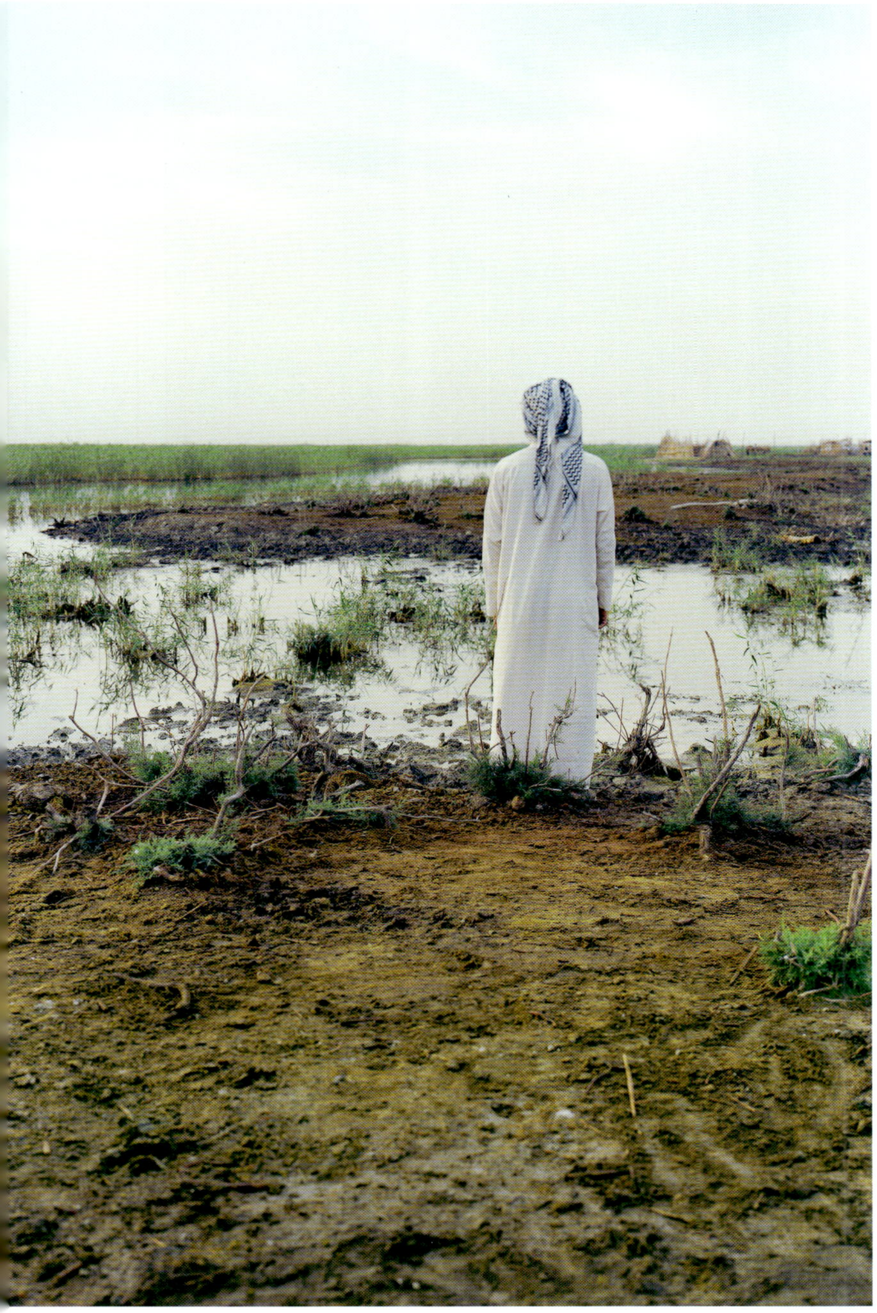

Ehmad and His Boat, Central Marshes, 2011–2012, 34 × 69 ⅜ in., linen (41 × [illegible]5 ½ in. with border)

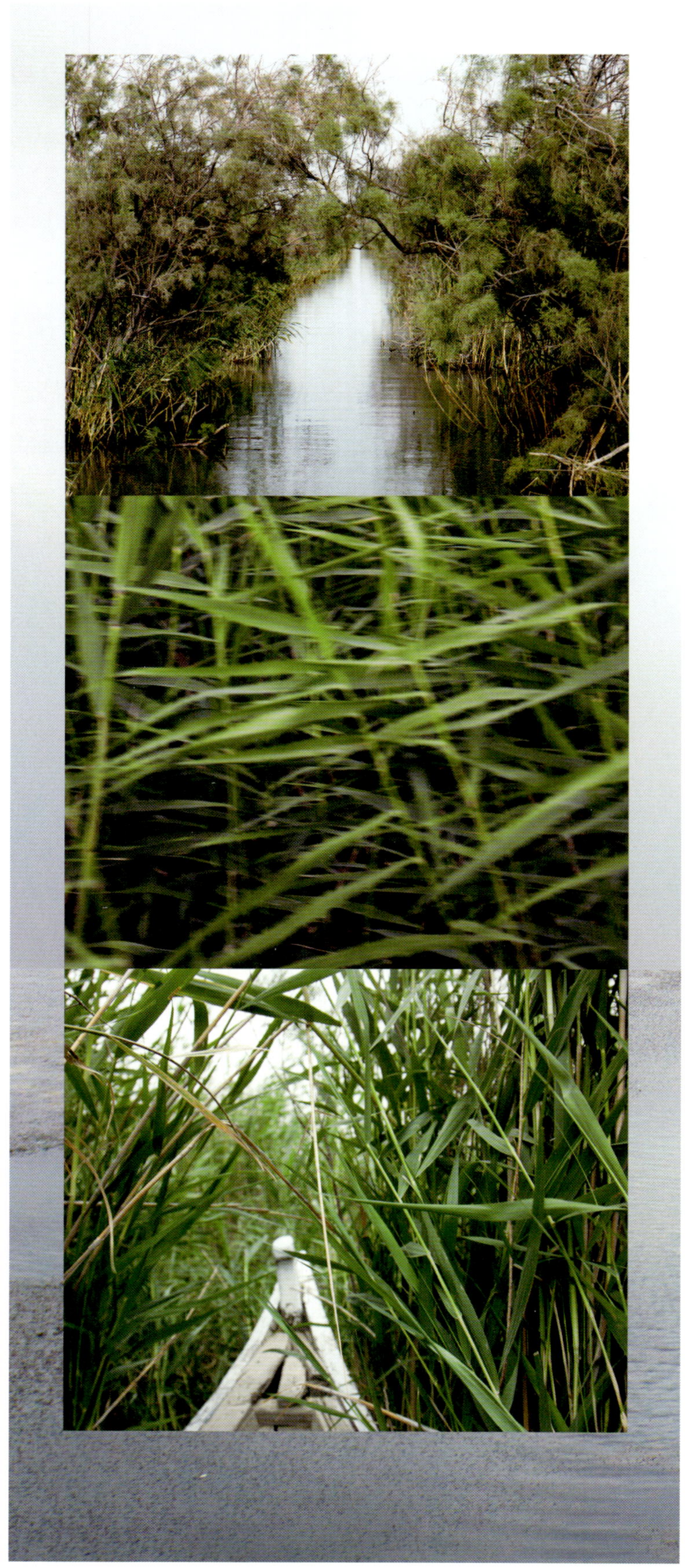

Eden Again, Southern Iraq Marshes, 2011–2017, 73 × 110 in., three hanging jacquard-woven tapestries

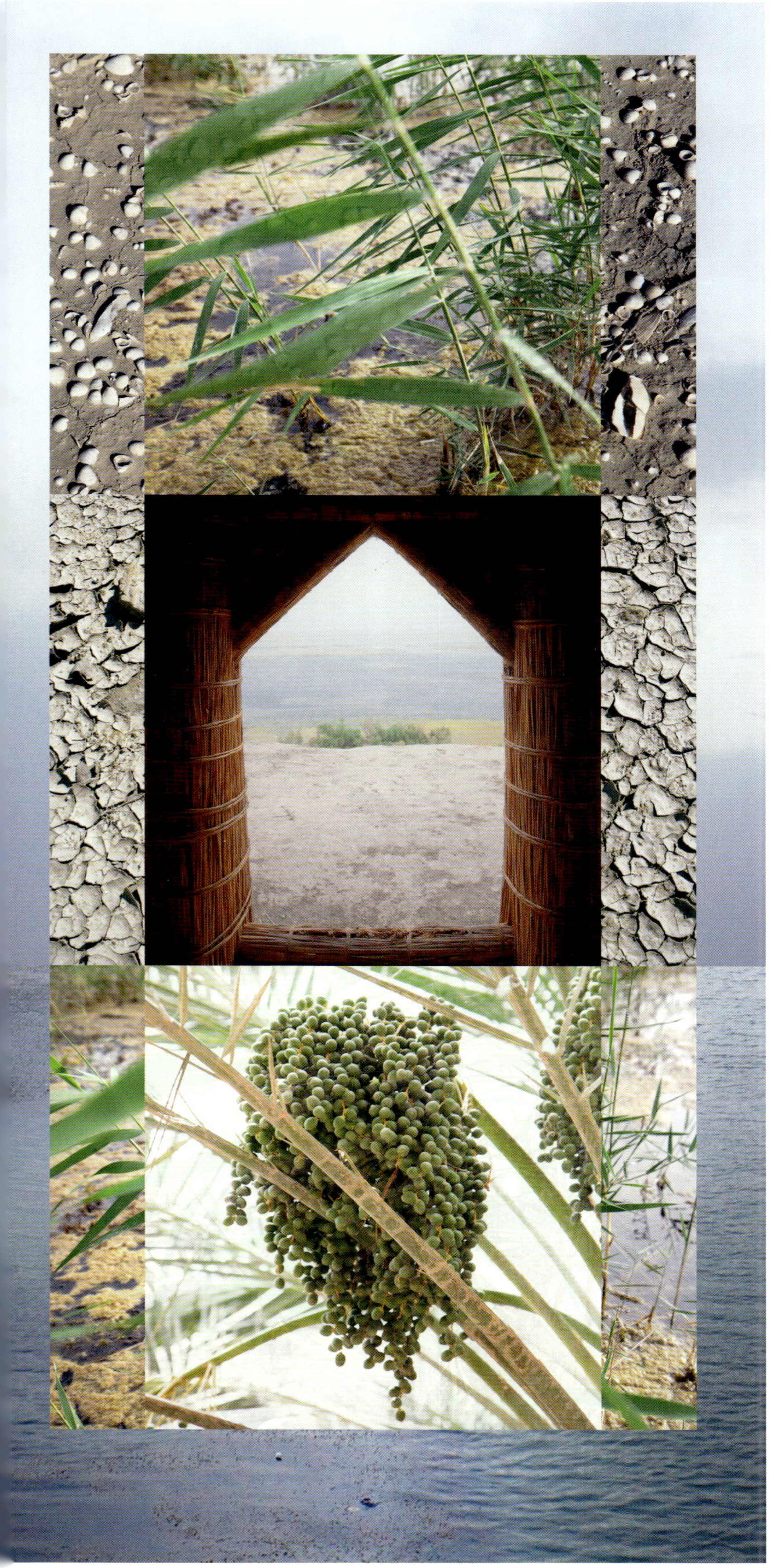

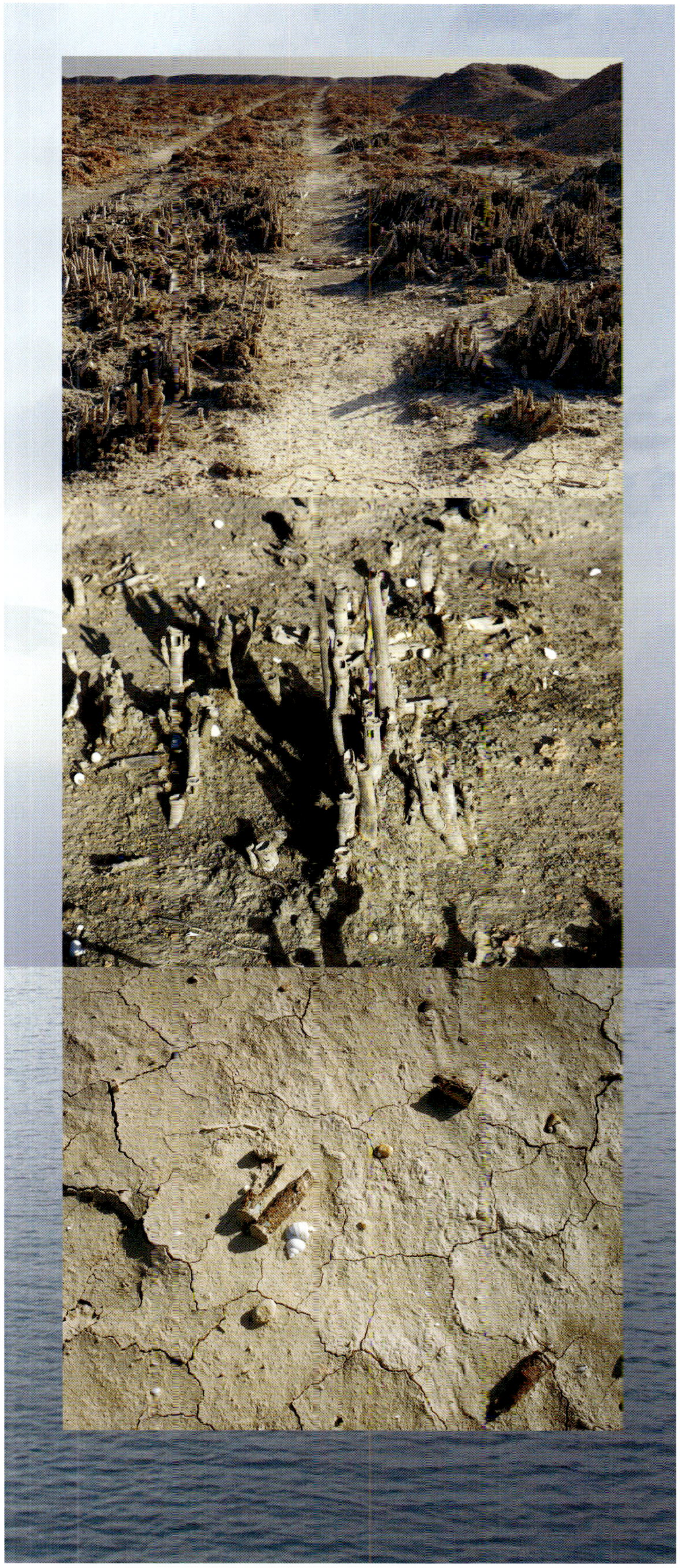

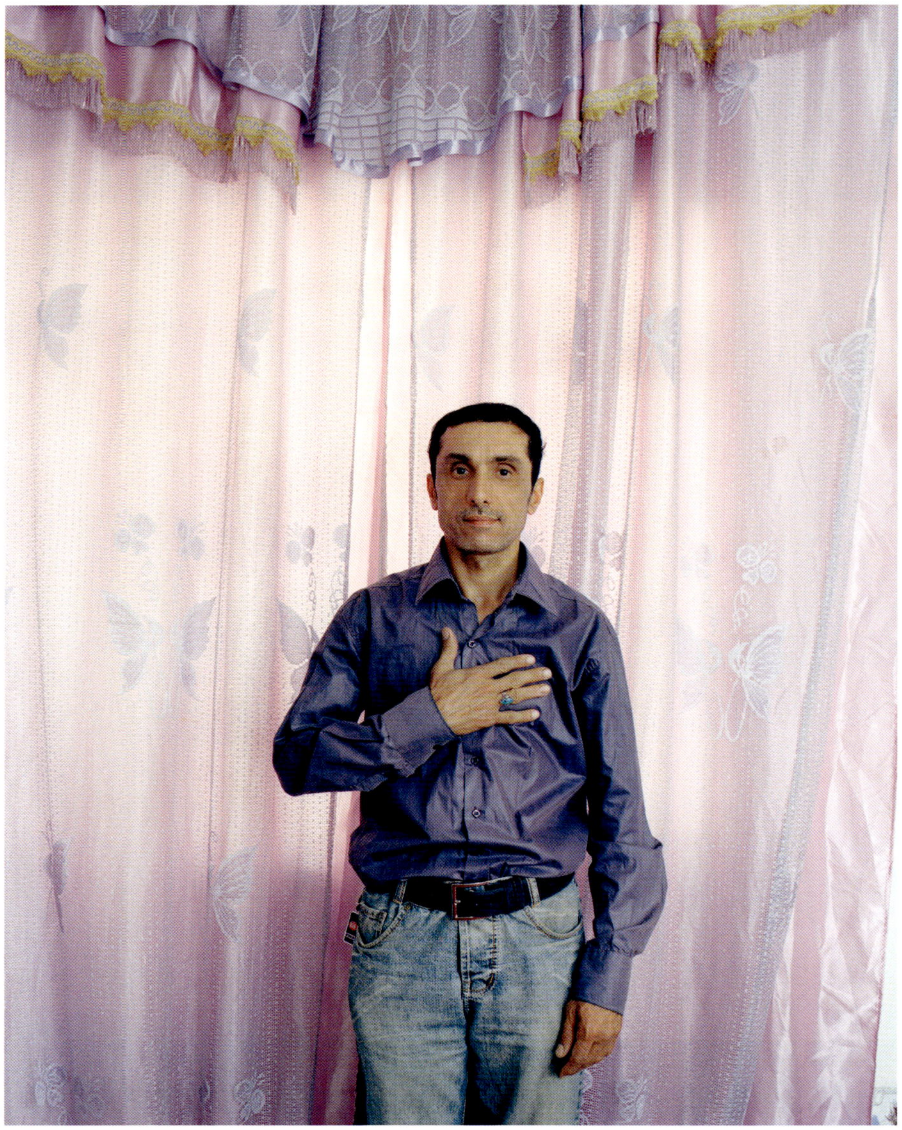

Najem Nassir, Al Fuhud, Southern Iraq, 2013

opposite: Amethyst Room, Southern Iraq Marshes,
2011–2016, 68 × 51 ½ in., linen (77 × 58 in. with border)

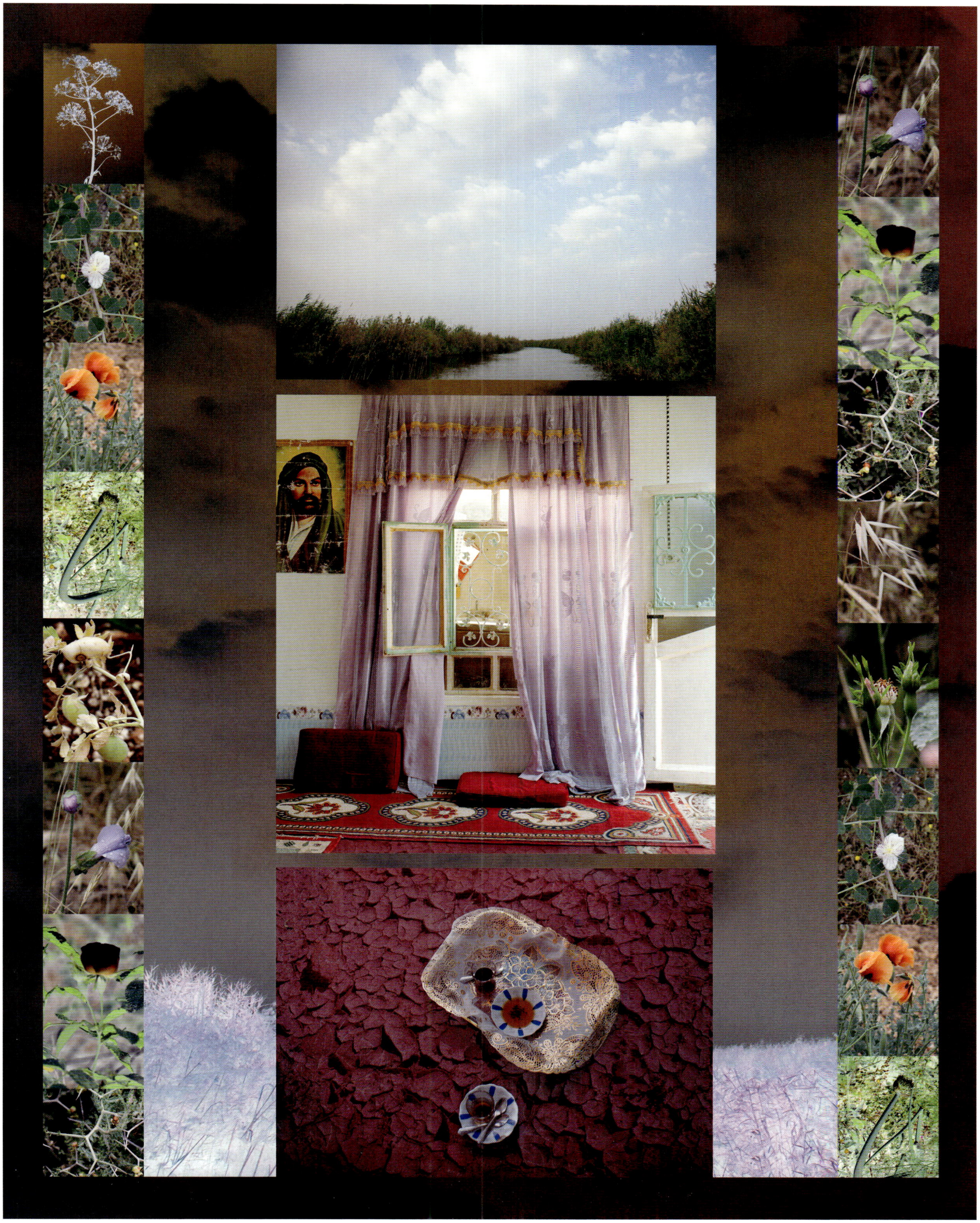

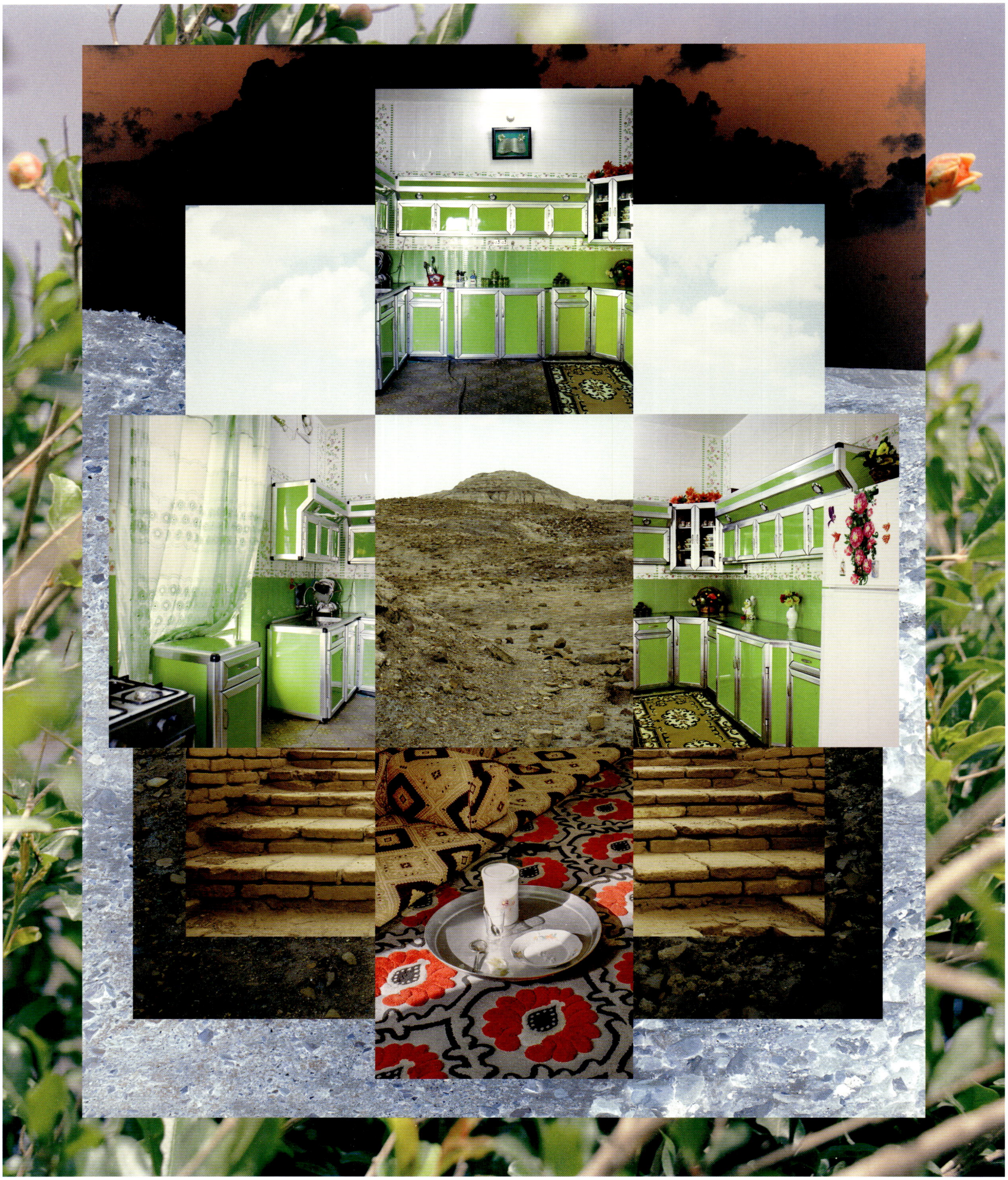

The Green Kitchen (with unexcavated Temple of Inanna at Uruk and steps of Ziguraut at Ur), Southern Iraq Marshes, 60 ¼ × 51 ½ in, linen. (69 × 58 in. with border)

Red Arch, Sayed Jusha, Southern Iraq, 2013

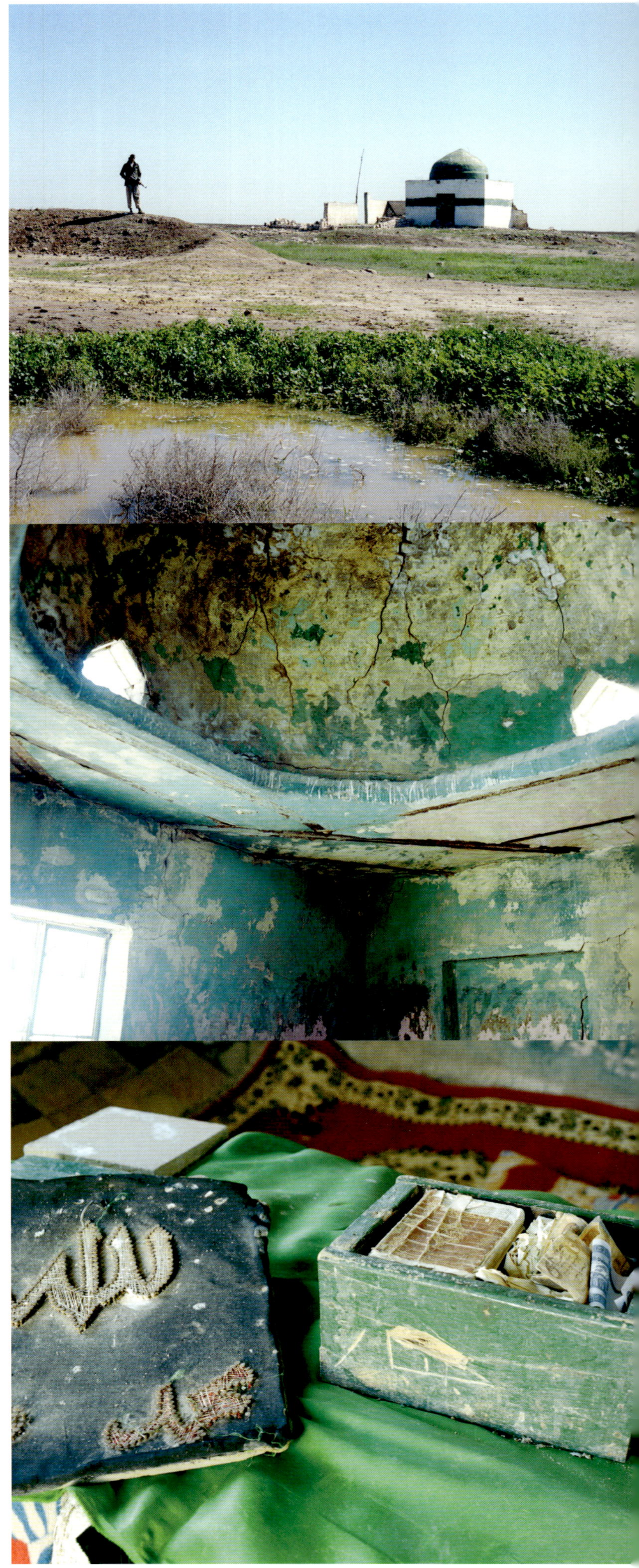

Saygal Shrine Palimpsest, 2011–2015, 38 × 48 in., aluminum

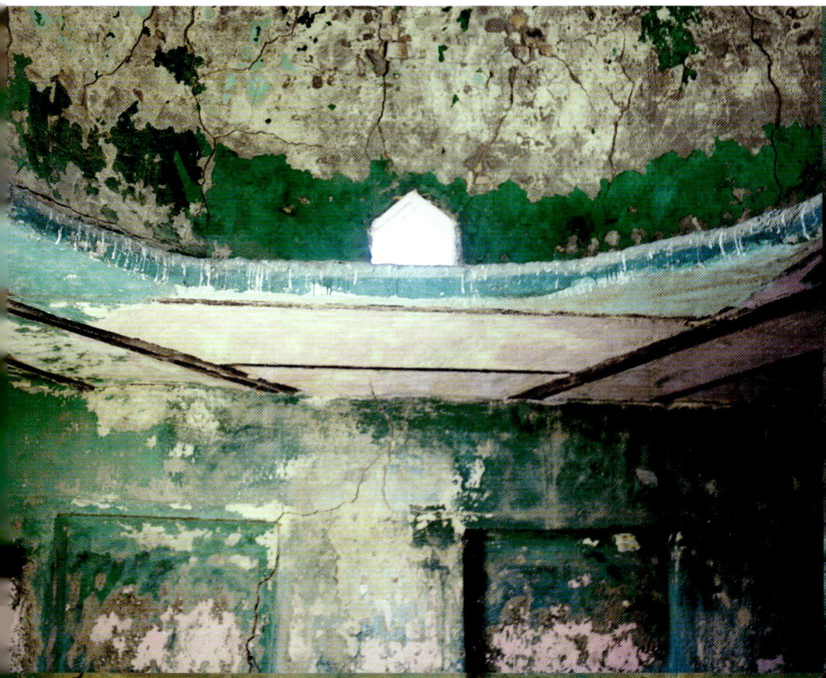
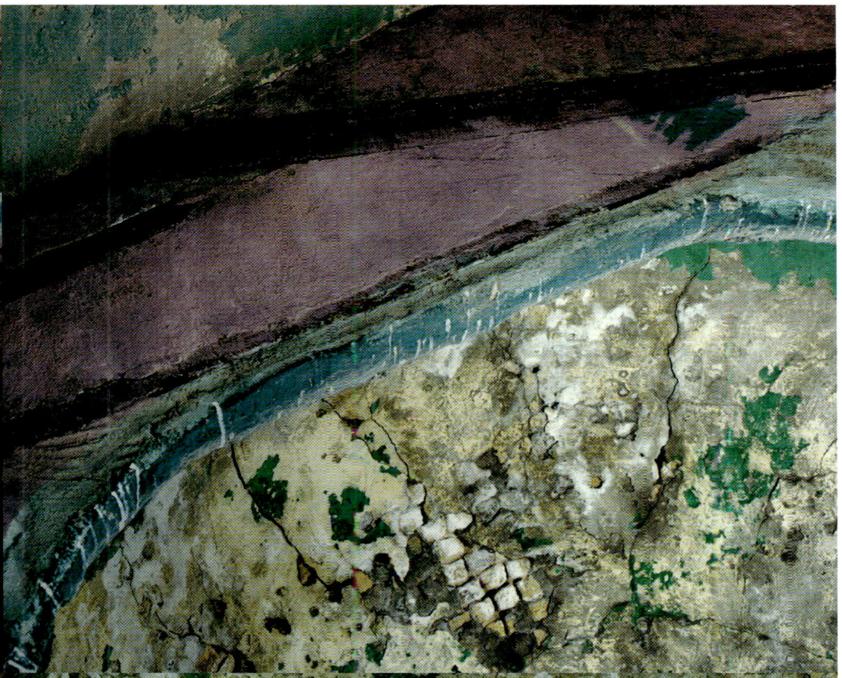
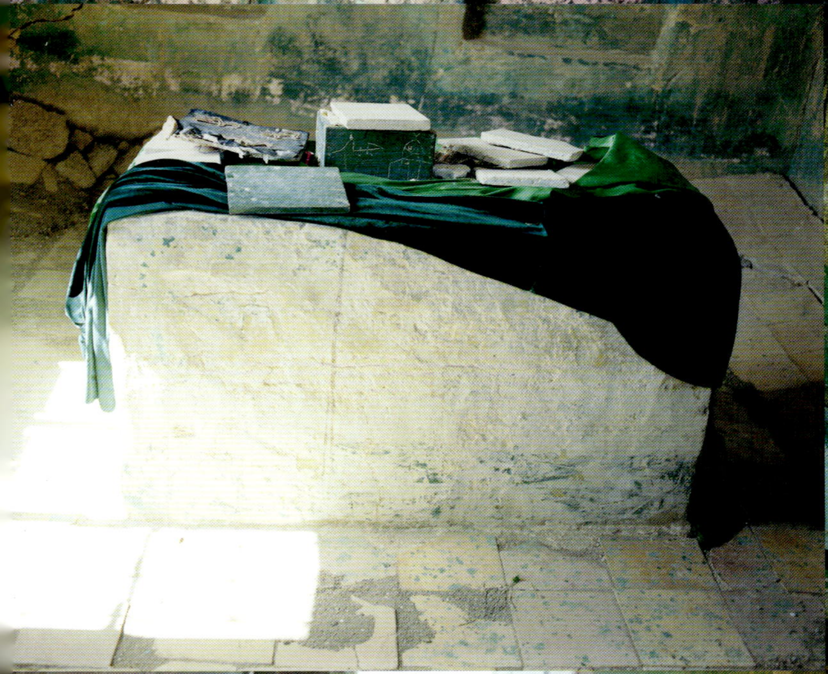
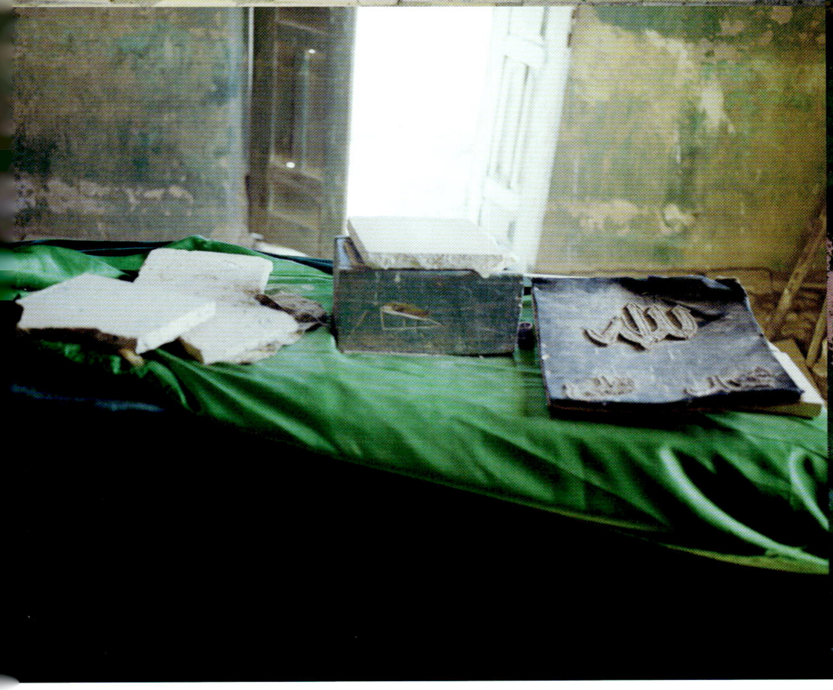
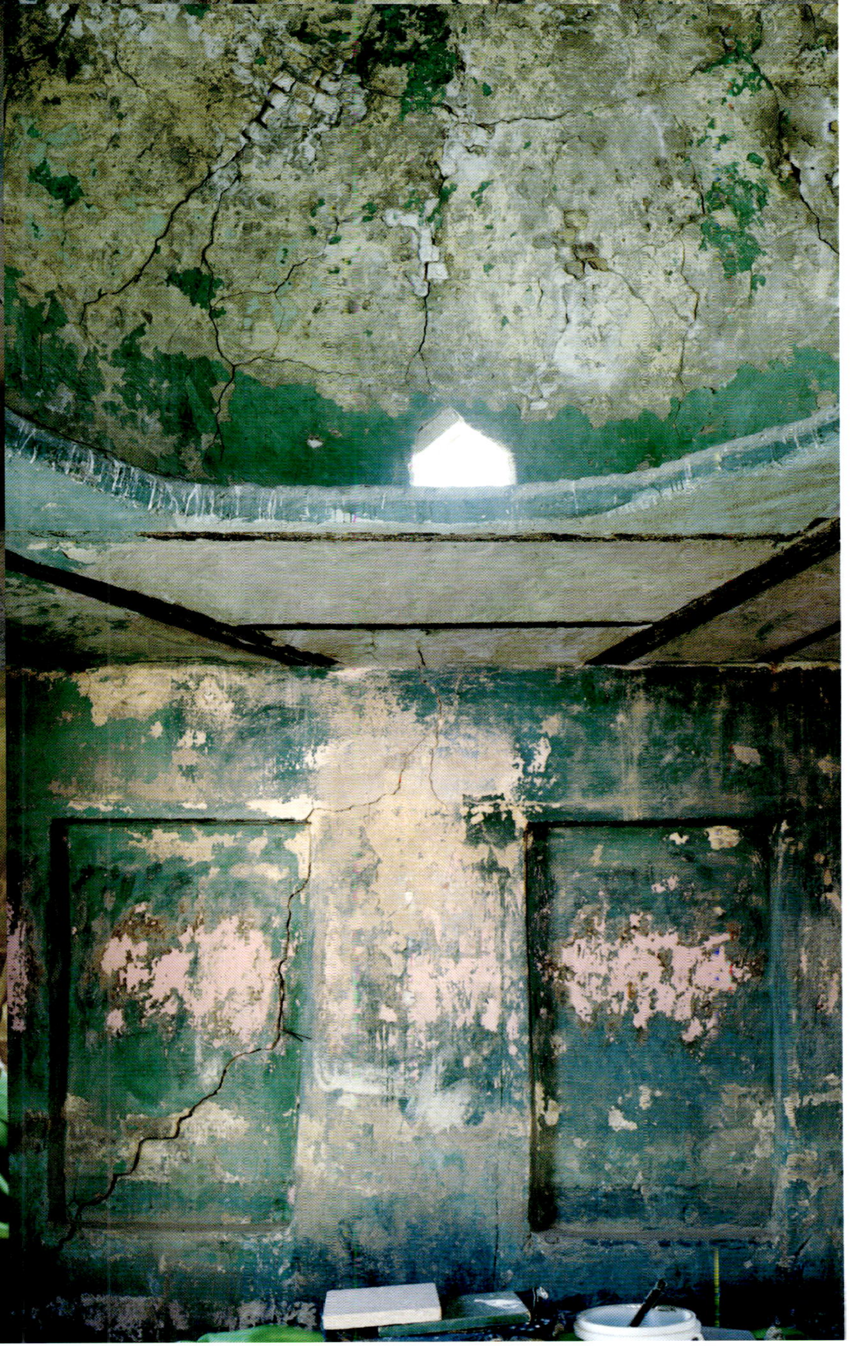

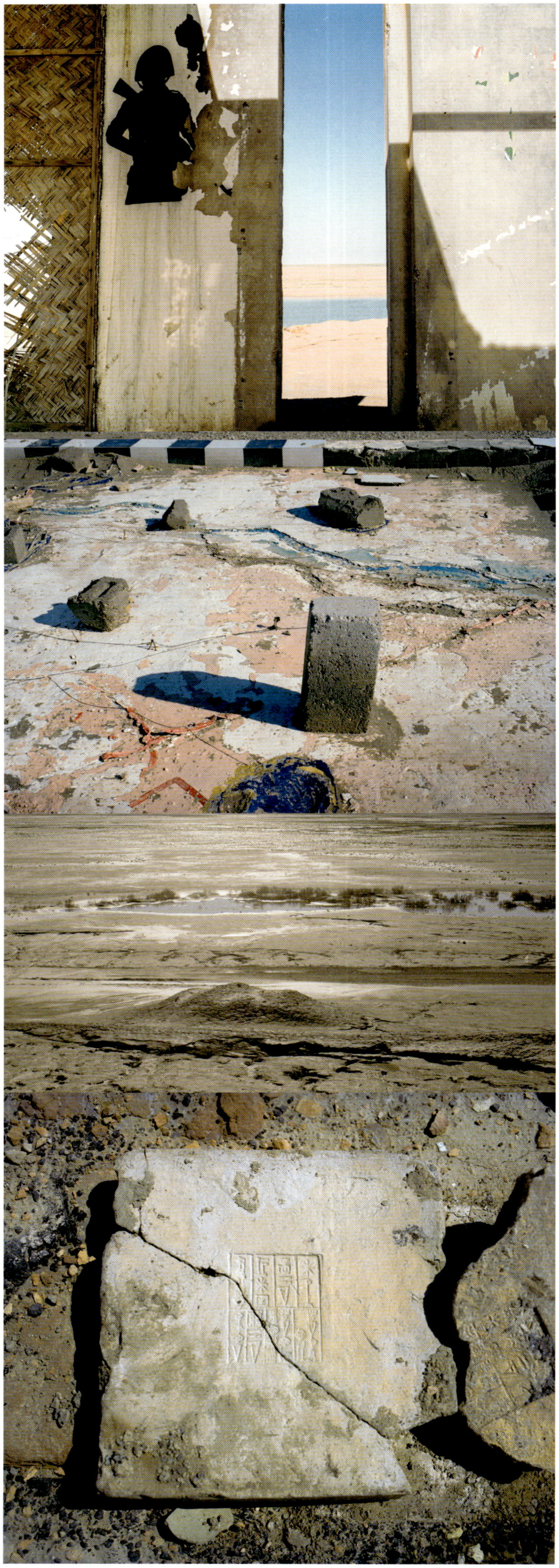

Military Site on the Way to Eridu, 76 × 27 in., aluminum

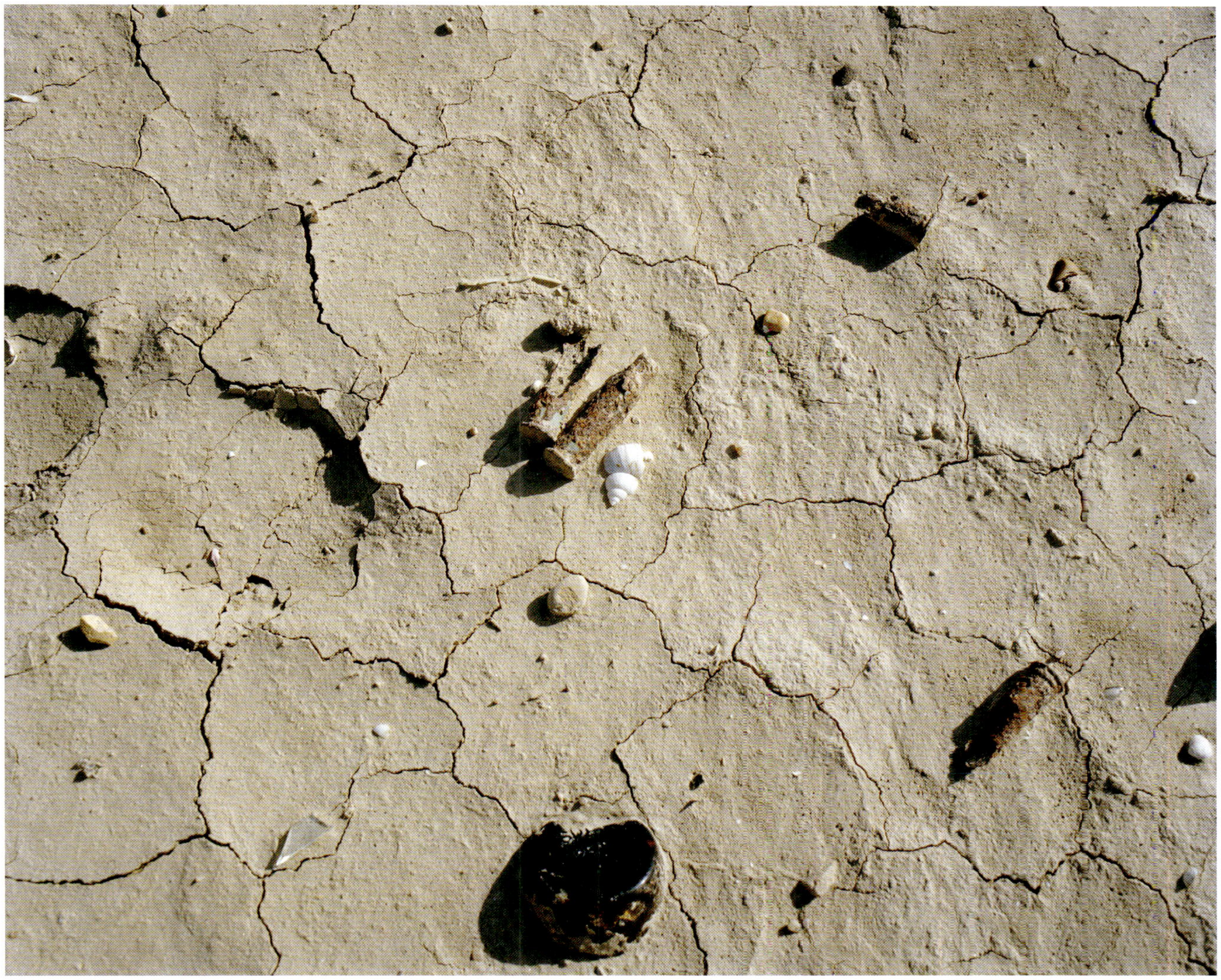

Shells and Shells, Saygal, Southern Iraq, 2011

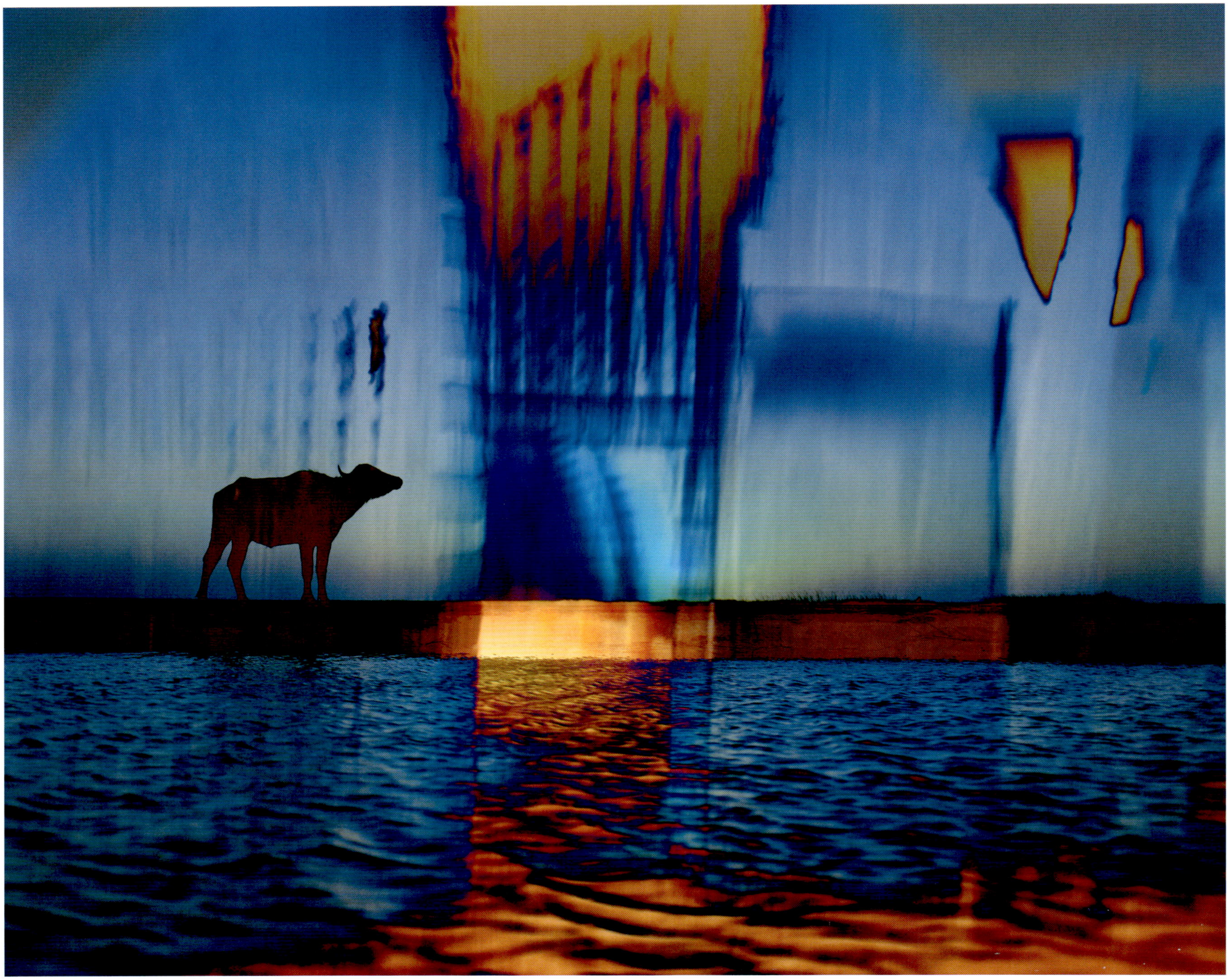

Dawn and Dusk: Mesopotamian Marshes, 2011–2013

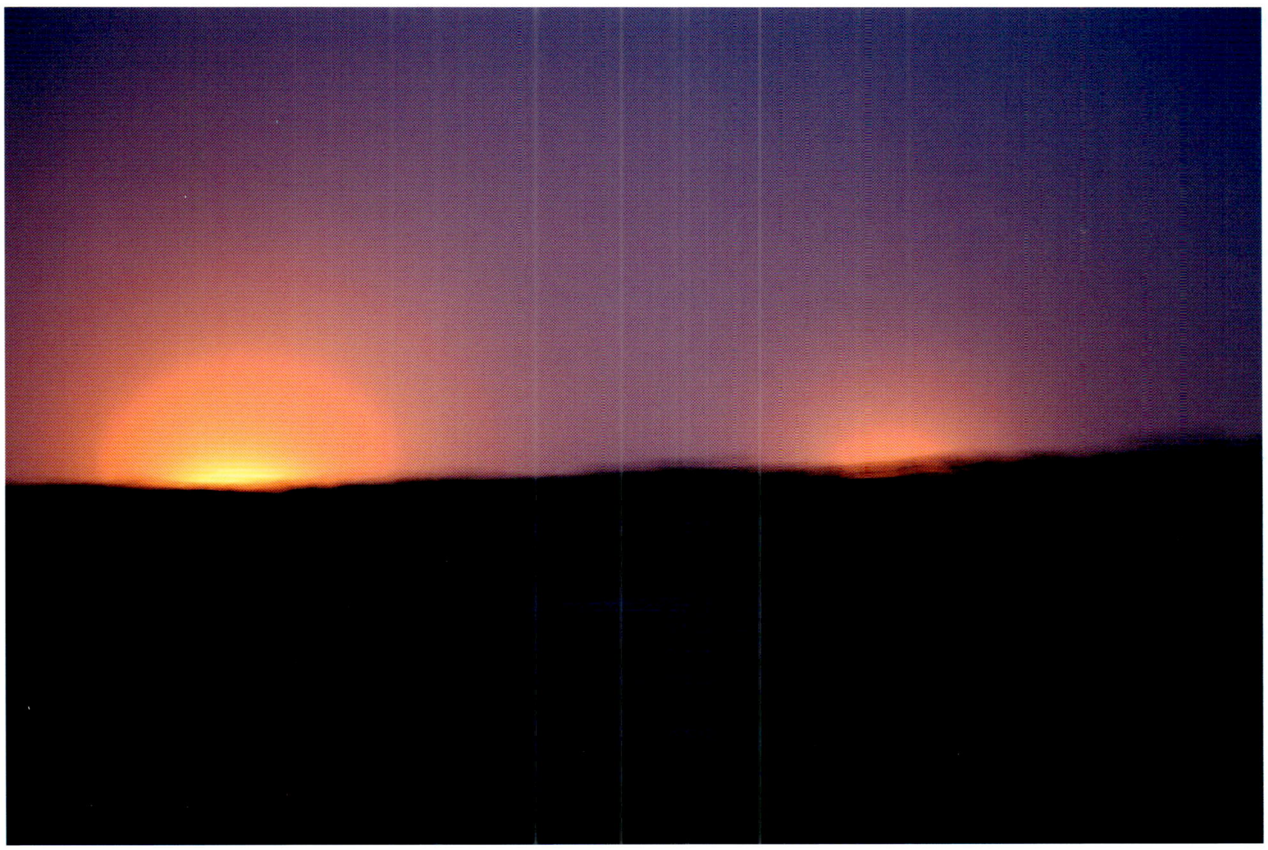

Double Sunset: Methane Flares on Highway from Al Querna, Southern Iraq, 2011

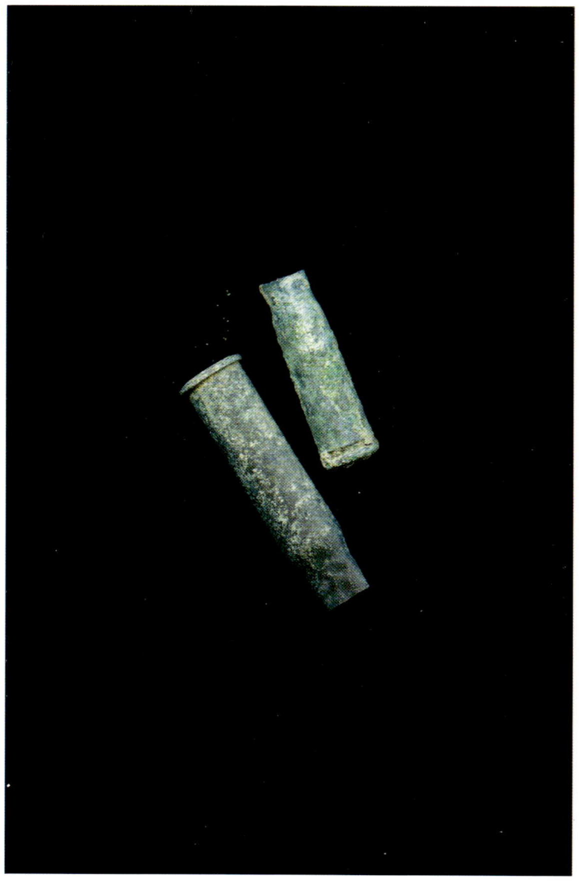

Temple of Inanna, with shells and shells, 2011–2017, aluminum; wall triptych, 72 × 91 in.; central column, 72 × 38 in., aluminum; shells, 33 × 21 ½ in.

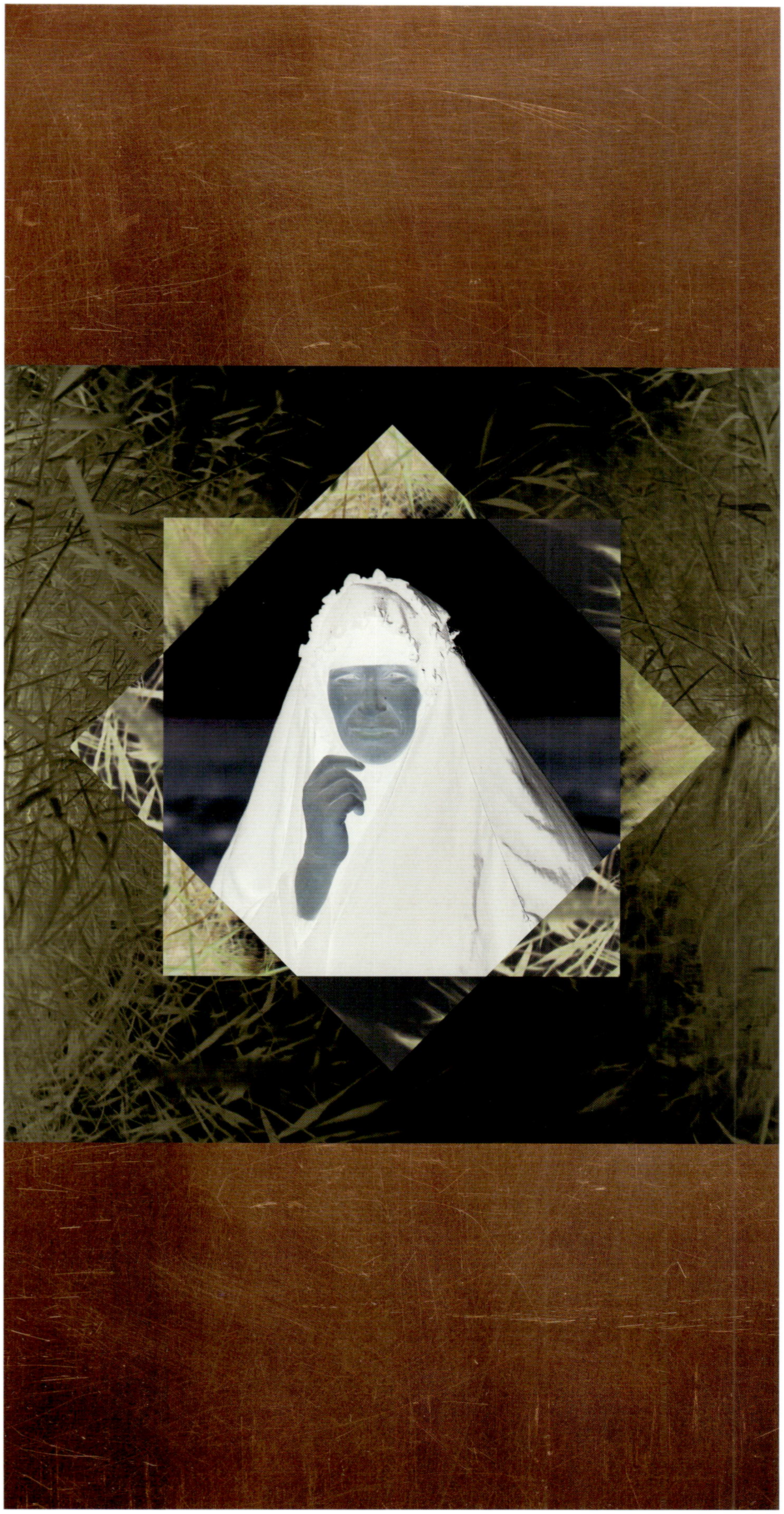

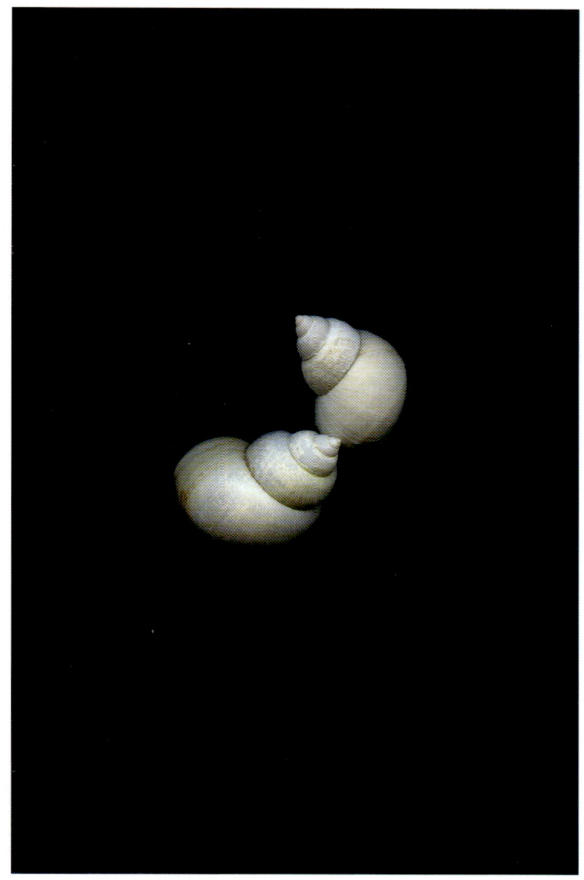

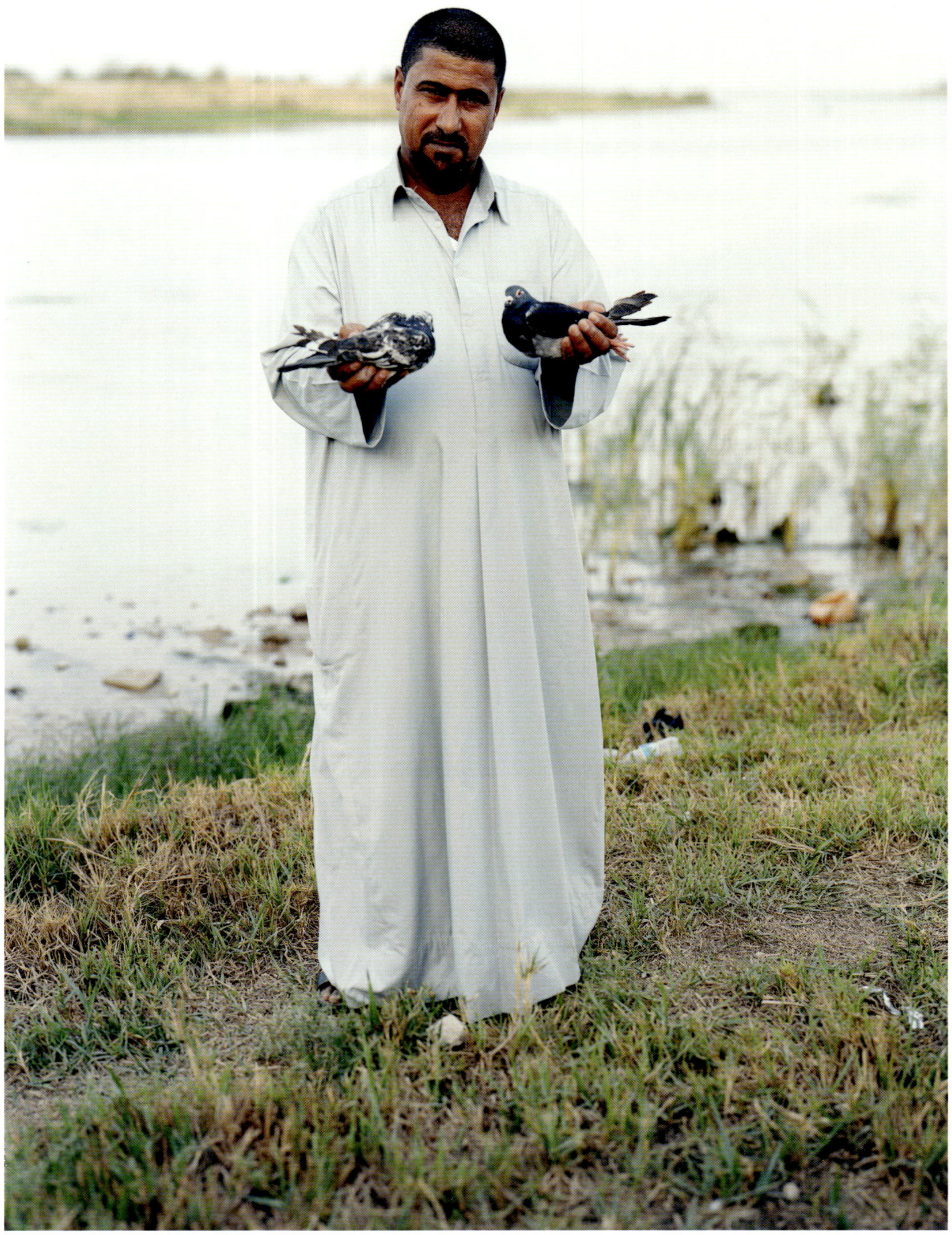

Hassan Jarry Al-Asadi with Doves, at the Shatt Al-Arab, 2011

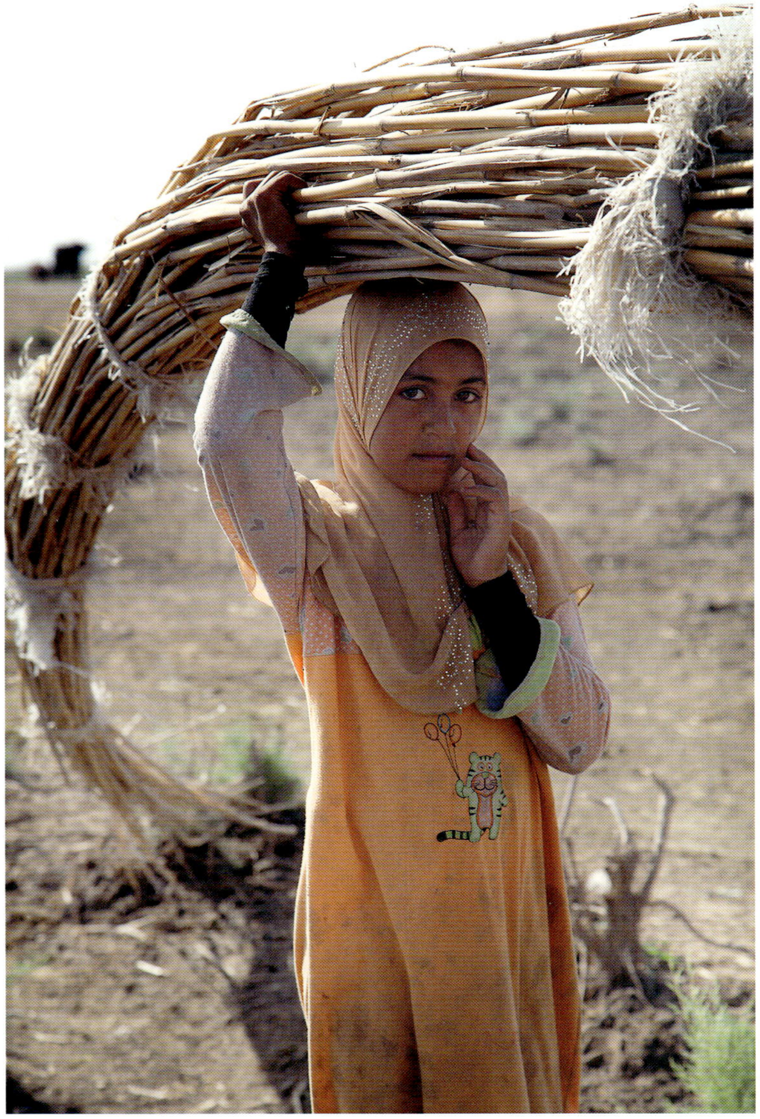

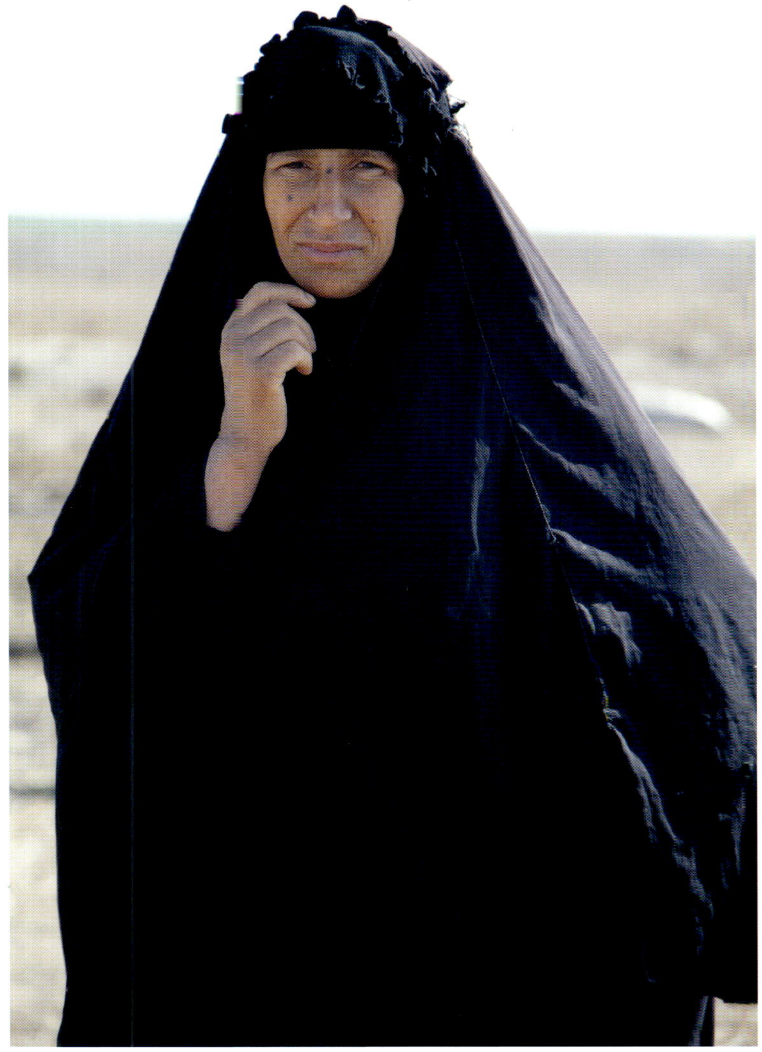

left: *Marsh Girl Moving House, Poisoned Water*, Central Marshes, Southern Iraq, 2011
right: *Marsh Woman Moving House, Poisoned Water*, Central Marshes, Southern Iraq, 2011

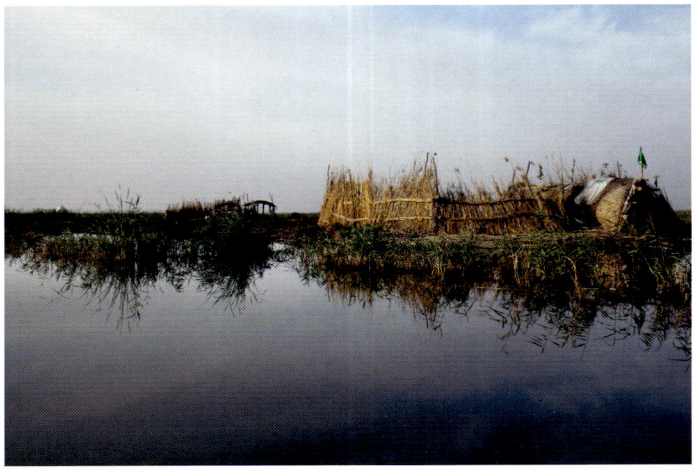
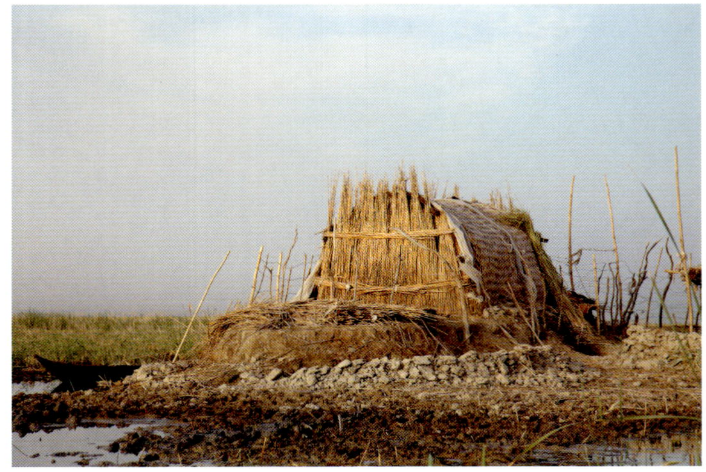
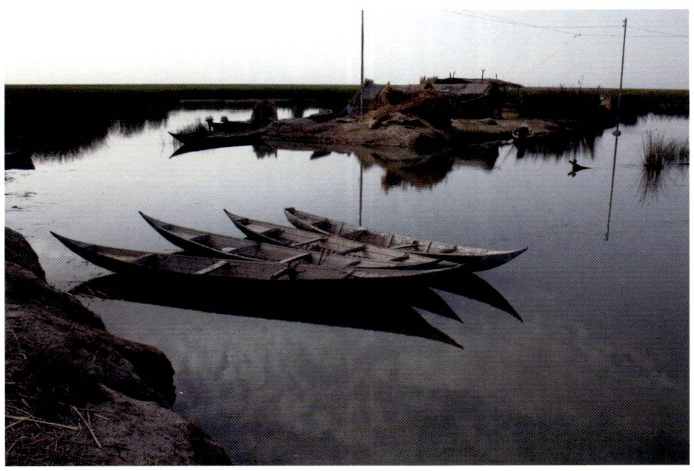
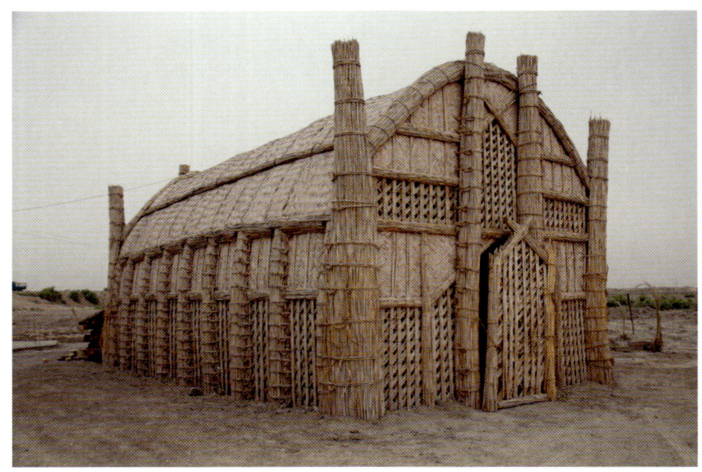
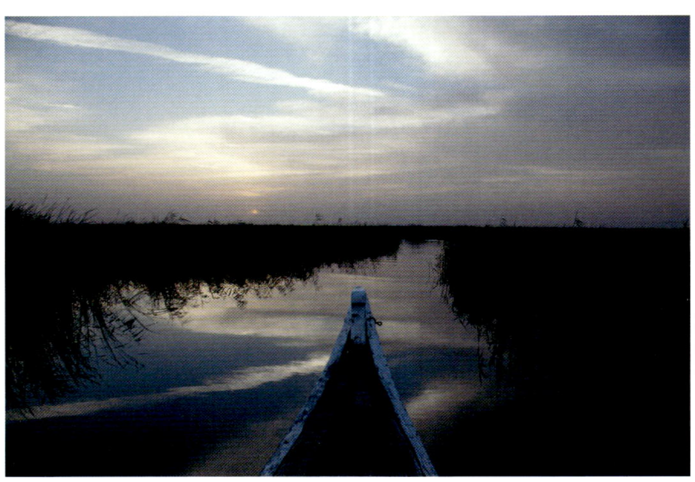
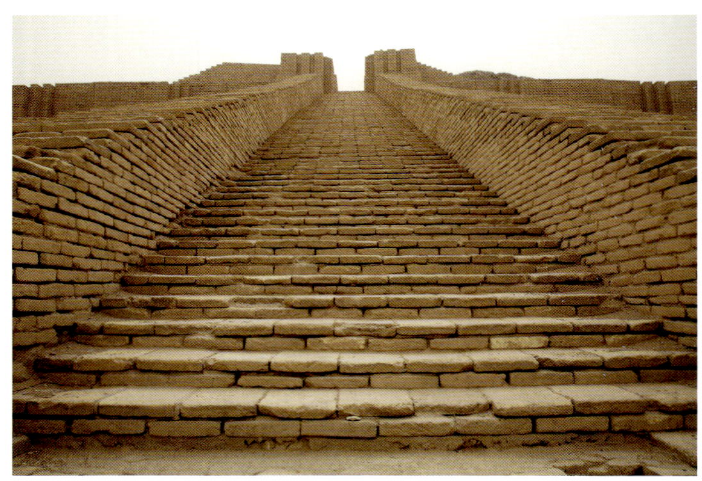

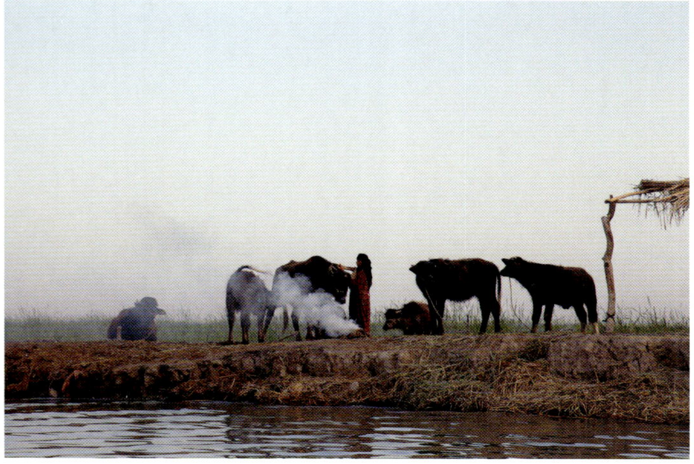

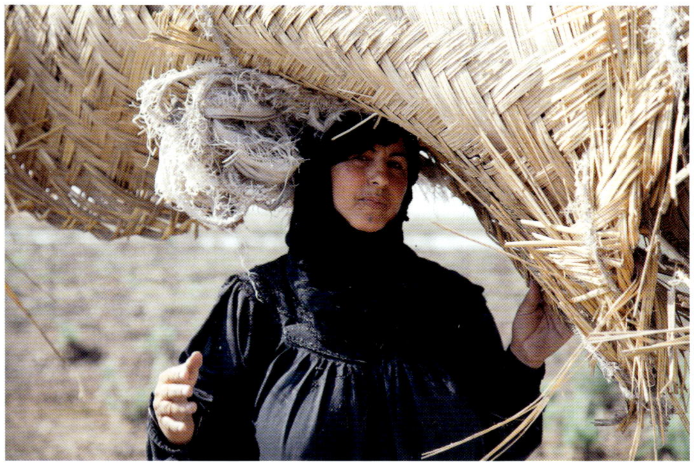

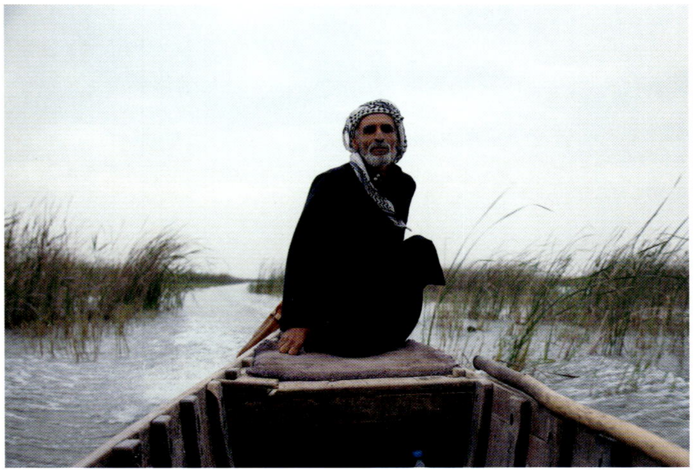

From the Southern Iraq Marshes, 2011–2014

Marsh Reed Compound with Green Flag, 2011

Buffalo Breeders Boats at Dusk, 2013

Ehmad's Boat at Dusk, 2011

Marsh Reed Hut, 2011

Nature Iraq NGO's new Mudhief, 2011

Steps, Ziggurat of Ur, 2011

Woman Smoking Buffalo to Rid Insects, 2011

Marsh Woman Carrying Reed House Rolled Up to Move, 2011

Abu Haider, Boatman and Musician, 2014

left to right:

Nelson Mandela's Cell Bars, Robben Island, Capetown, South Africa, 2006

Wangari Maathi, Founder, Green Belt Movement, Nairobi, Kenya, 2006; in Boston, Massachusetts

Freedom Fighters Forest, Mt. Kenya, Africa, 2006

Irish Soda Bread, 2006

Bernadette Invoking Saint Goubnait, Southwest Ireland, 2006

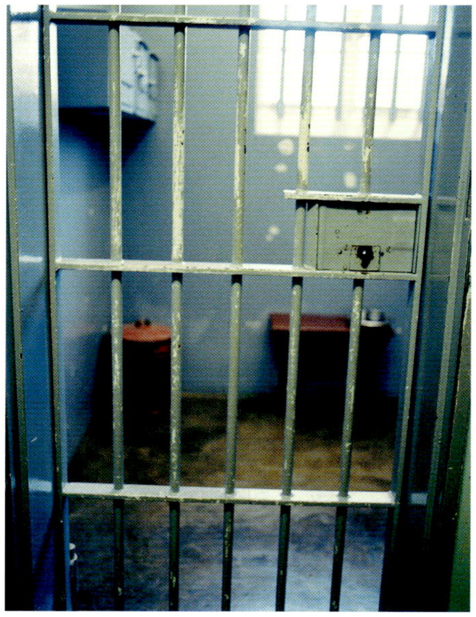

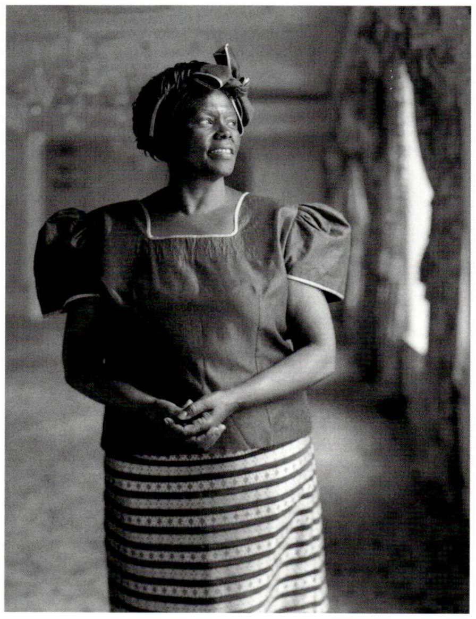

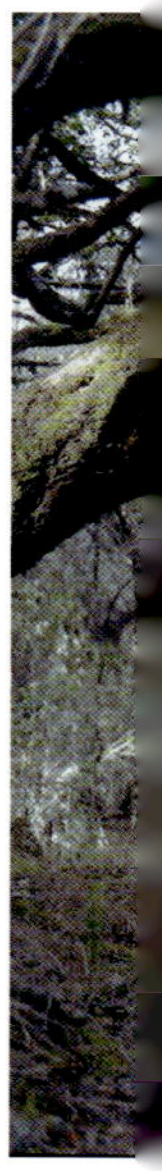

Trilogy

Meridel Rubenstein

Like most of us in this culture, I used to see the world in dichotomies and oppositions (nature/culture, city/country, etc.). Making triptychs is one way I found as an artist to transcend polarities and have a better view of the multiplicity of the whole. With this three-part work, *Eden Turned on its Side*, I wish to transcend specific locations and view the planet as one garden. Nevertheless, this project is embedded in the ecologies of the places I call home—Vermont, New Mexico, and Singapore (where I teach)—and in the Southern Iraq marshes near the historic site of the Garden of Eden, which you might say is an ancestral home for me.

Before I began this project, I made a "walkabout" in 2006 to explore different ecologies and earth guardians. This led me first to East Africa to visit three people I greatly admired. The first was a colleague from my year at the Bunting Institute at Harvard (1998–1999), psychologist Dr. Pumla Gobodo-Madekizela in Capetown, South Africa. Pumla had been a member of the Truth and Reconciliation Commission. We had an idea for a Public Art Project at Robben Island where I visited Nelson Mandela's prison cell, now a memorial site.

Next I visited my cousin Steve Goodman, a renowned wildlife biologist and MacArthur Fellow, who was working on a project to save bats from extinction throughout islands in the Indian Ocean. I spent a month with him in the wilds of Madagascar, trapping bats.

Lastly I volunteered as a photographer for Nobel Laureate Wangari Maathai and the Green Belt movement in Kenya, documenting their communities and tree nurseries. I wanted to experience firsthand humanitarian and ecological work outside of the art world where I had been immersed for over thirty years. A few months later, I had a month-long artist's residency at the Sirius Arts Center in Cobh, Ireland, where I continued my study of ecology, this time learning about Ireland's wildest places as refuges for soldiers healing from combat.

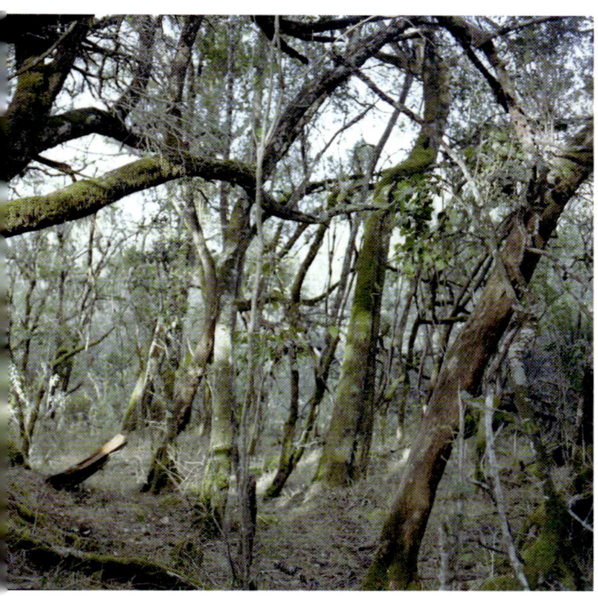

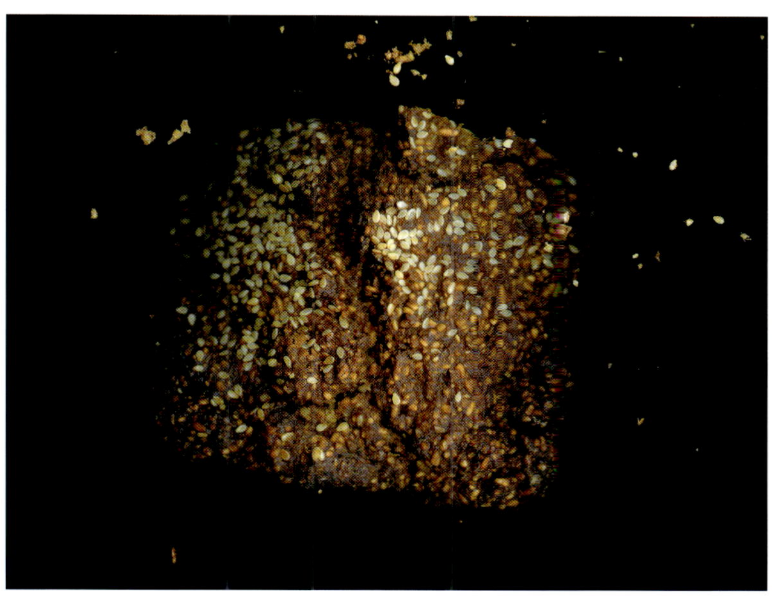

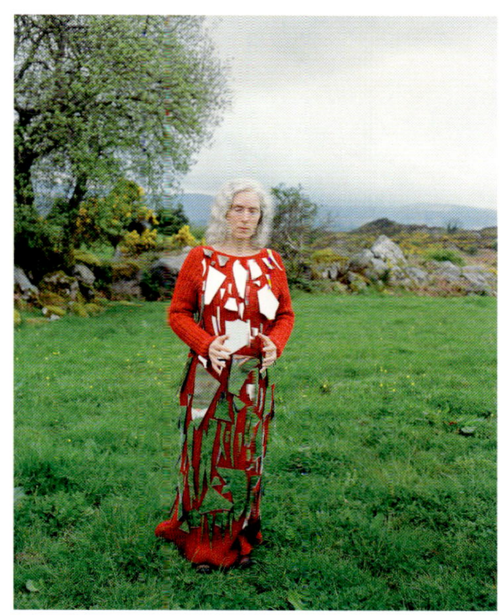

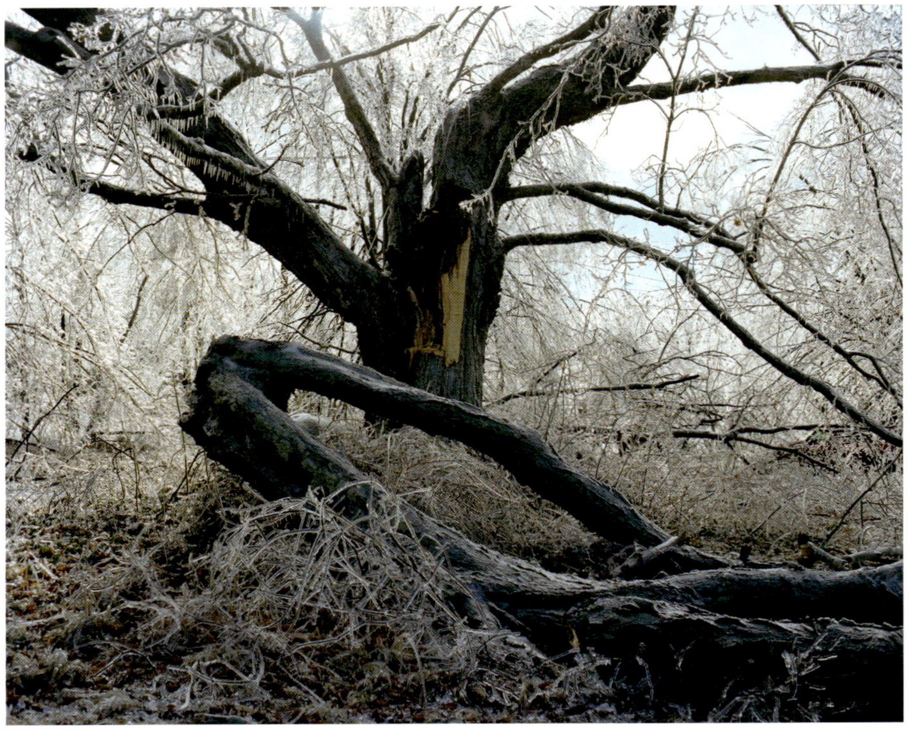

Ice Storm, Broken Maple, Vermont, 2008

Photosynthesis

One late evening in December 2008, I was alone with my dog Kita in the barn cottage of our family farm in Southern Vermont. Though I was born and partly raised in Detroit, I was inseparable from this land even after thirty-five years in New Mexico. I had just returned from teaching in Singapore. An ice storm quickly blew in, and ancient broken tree limbs were crashing down on the barn's tin roof. The effect was deafening and reminded me of the sound of armaments. I knew the power would soon go out, so I grabbed Kita and drove down the hill into what looked like a war zone. I was probably the only car to get out for the next few days, until the roads would be cleared of fallen trees and hot electrical wires. But when I did return, the farm was a wondrous glazed firmament of crystalline trees and archways. One of the most ancient maples had lost a huge limb. I immediately went to the local elementary school and invited the science class to come over to figure out how to save the tree by the miracle of photosynthesis, in the dead of winter.

In *Photosynthesis* intersections of nature and culture are presented in relation to the biosphere and its seasons. For several years, at each solstice and equinox, I photographed people and trees, portraying them as trying to support each other. Is it possible to restore balance between oxygen and carbon, between humans and the natural world? Photosynthesis is the predominant light-recording system that sustains life on earth. A by-product of this process is respiration, whereby plants exhale oxygen that we anthropoids inhale while exhaling C02 for plants in return. This Edenic symbiotic exchange is the subject of this work. In Vermont they drill holes in maple trees and install tubing to draw the sap into distilleries to boil down into maple syrup. I hooked up the sap lines from the Vermont maple trees to oxygen masks to illustrate how people and trees exchange carbon and oxygen. We breathe in and out, keeping each other alive. On the monitor in the studio, I drew a circle on a cloud image and then transformed the image to its negative. The opposite of sky is solid earth. Out of nothing, something. To emphasize the feeling of the atmosphere, the lightweight paper prints float in the frame, suggesting the lightness of oxygen molecules.

Volcano Cycle

We are in East Java making our way up a three-thousand-foot mountain in order to hike back down another three thousand feet to the Kawah Ijen Crater lake. I am hoping to find a volcanic site that resembles the primordial Earth. The first view from above of the steaming crater lake takes my breath away. The yellow gases and the extreme aquamarine of the lake do look like the beginning of time. But we have to keep going because clouds are moving in. Workers are carrying heavily loaded twin baskets balanced on a pole across their shoulders uphill. These contain hardened sulfur congealed from the gases emanating from the crater, which is a combination of Sulfur and Hydrochloric acid. We get to the bottom and, as I set up the view camera, the clouds burst. The workers help hold the tarp over the camera while I make only one exposure. Then we run to a little crevice in the rock canyon walls to wait out the rain. After a few hours the rain hasn't slowed down at all, and the workers' lips are turning blue. Rivers are pouring over the sides of the canyon, and soon our trail up will be submerged. We make a dash for the trail. By the time we reach the top, hours later, the rain has stopped and the Javanese air warms us. The one exposure is the best of the volcano images.

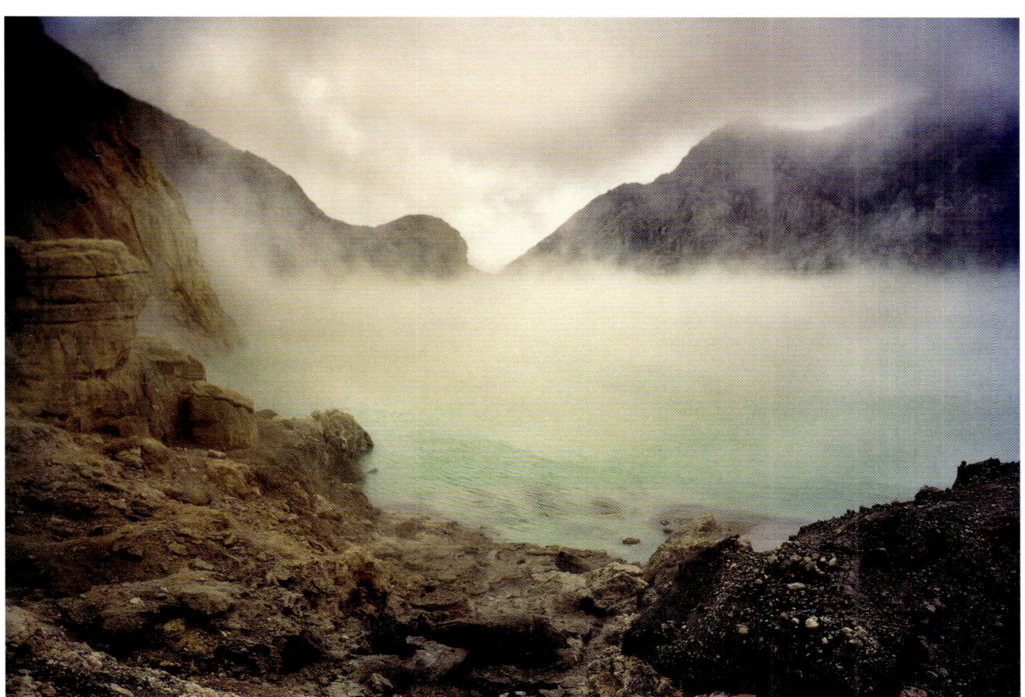

Acid Crater Lake, Kawah Ijen Volcano, East Java, Indonesia, 2011, 43 ¾ × 30 in.

Dan Schrag and Paul Hoffman's hypothesis "Snowball Earth" theorized that the most severe ice age of the last billion years occurred from 850 to 630 million years ago and encased the earth in ice. A thaw resulted from volcanic activity that trapped an accumulation of CO_2 (greenhouse gases) in the atmosphere, causing global warming. My artist's mind envisioned the released oxygen and water flow spurring the birth of zillions of multi-cellular cyanobacteria, beginning to breathe us into being along the evolutionary path. Those early volcanoes produced another kind of Eden.

Mt. Toba in Sumatra, Indonesia, erupted seventy-four thousand years ago, creating a volcanic winter that allowed only a few humans to remain; those with the ability to make tools could climb out of the ash. This produced a bottleneck of human evolution, reducing the global population. A six-year global winter reportedly ensued. Toba is called "The Weak Garden of Eden" because creation and evolution had to begin again. The earliest volcanic eruptions allowed for global warming that released sufficient carbon dioxide and oxygen to begin life on earth. The study and imaging of volcanic cycles allows an immediate recognition of the power, fragility, endurance, and persistence of life on earth.

My experiences with the inhabitants of the volcanic landscapes of Mt. Merapi, Mt. Bromo, Mt. Batur, and Kawah Ijen enabled me to see the coexistence of nature and culture both in and out of balance. Volcanic soil is among the richest on earth, so people do not want to leave their farms. Twentieth-century volcanology has led to very successful education programs and warning systems for those who choose to remain, as in the vicinity of Mt. Merapi, which can erupt every four to six years. In *Pyroclastic Remains*, Mt Merapi, photographed after the last eruption in 2011, is superimposed upon itself before the eruption in 2010.

I chose to print the images on prepared aluminum plates. The metal echoes the deep timbre of the eruptions, as well as the transmutation of materials that occurs during a volcanic event—with exploding glass shards and running rivers of melting metals. The metal is resonant of deep time that has passed to bring us into being.

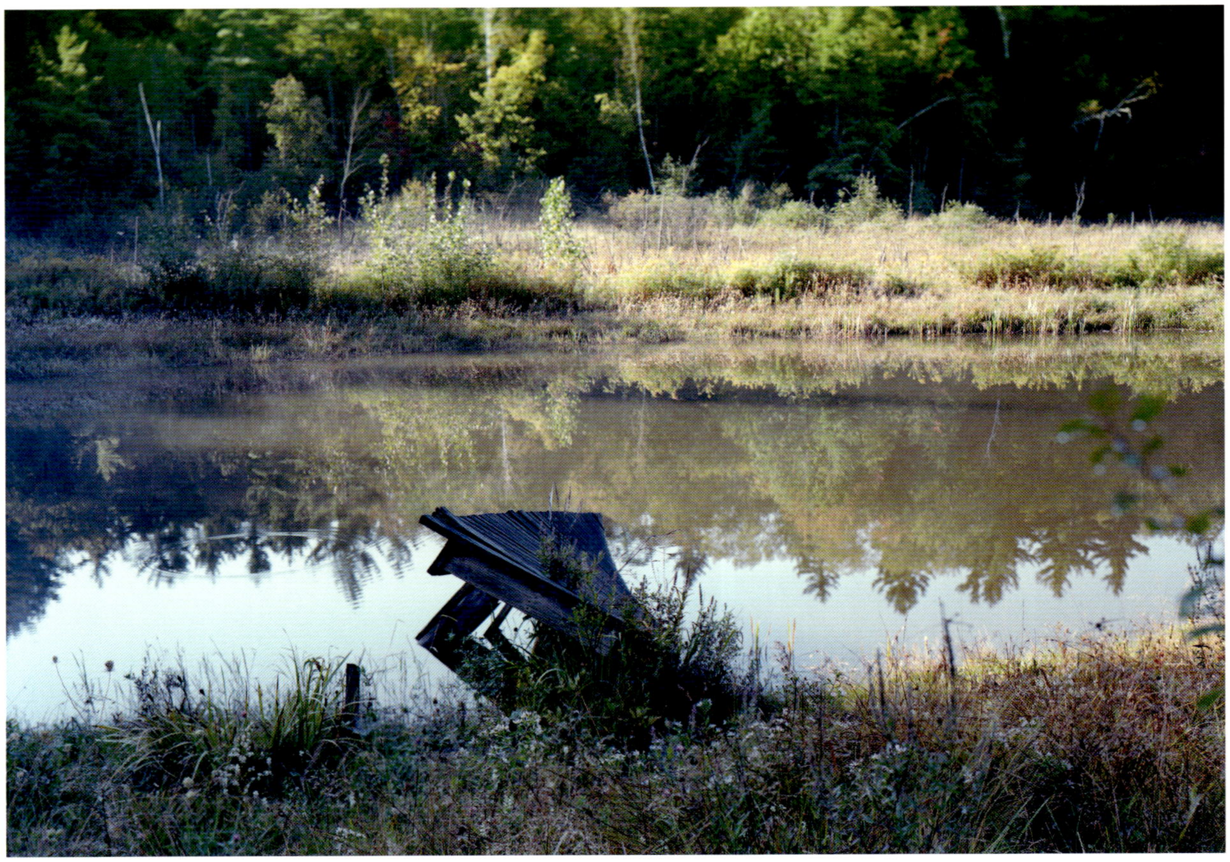

Farm Wetland, Vermont, 2012

Eden in Iraq

In September 2006, after a month in Singapore as the first Art and Technology Visiting Artist at the School of Art, Design, and Media, I return to the Vermont farm. I'm awakened at 3:30 a.m. by the howling of coyotes circling the newly destroyed wetland/beaver pond. They are moving in for the kill. My neighbor had been contesting the few feet this wetland encroached on her land. She used the legal excuse of a few trees to log to disturb this wetland. On a stealth mission six nights before I return, the dike is breached and the wetlands drained. I've returned just in time to see the remains of the destroyed wetlands, where wild grasses, beaver, geese, ducks, heron, bass, turtles, frogs, otters, and muskrats have made this area a paradise. My friend Dickie has made daily jaunts there to feed the birds and watch the evolution of this swamp into an oasis. The wetlands were hastily drained, sending acres of muddy water, sediment, and alien bass into the town's main (trout) stream and water supply. As soon as the level was down the coyotes appeared, and after six days the area is a wasteland.

In 2010, in the midst of creating both *Photosynthesis* and *Volcano Cycle* for *Eden Turned on Its Side*, I was alerted that the television series *60 Minutes* was airing a program called "Eden Again." It featured the valiant efforts of the Iraqi Environmental NGO, Nature Iraq, to restore the vast drained Mesopotamian marshes of southern Iraq. Home to the Marsh Arabs, descendants of the first settlers of nearby semi-excavated Eridu, Uruk, and Ur, they had lived an unchanged, water-based way of life here for over seven thousand years, with their beautiful woven-reed architecture often built on floating reed islands, colorful embroidered wedding blankets and water buffalo trade. Saddam Hussein drained the wetlands to punish Shia rebels hiding there in the early 1990s. Like my Vermont neighbor on a stealth mission, he built a secret road into the third largest wetlands in the world and the largest in the Middle East. His engineers then constructed a giant earthen canal to secretly empty the waters of the Tigris and Euphrates rivers (which flank and fill the marshes) into the Persian Gulf. These rivers cross on the eastern edge of the marshes so their waters are thought to be in extreme proximity to the original historic Garden of Eden. When the waters got low enough, Saddam sent in tanks with mortar and fire to burn and destroy, turning Eden into a scorched desert. With thousands killed, all but a few of the surviving 250,000 Marsh Arabs fled to nearby cities, to Iran, and to the suburbs of Detroit and San Diego. With the demise of Saddam Hussein in 2003, people started returning, with simple hand tools, to break through the earthen canal to let the waters back into the desiccated land. Azzam Alwash, an engineer from Laguna Beach, California, immediately returned to the nature preserve of his childhood. He created Nature Iraq with the mandate to restore and preserve Iraq's endangered waters, flora, and fauna. I wrote to Azzam. He invited me to come.

Here was a living example of a vanquished people returning to regreen their Eden. The restoration of this land and culture is a brilliant model of resilience and sustainability. But there were problems in Eden. Because returning Marsh Arabs were now building on land instead of water, the Euphrates was seriously polluted with human waste and had become a health hazard. I initiated the *Eden in Iraq* project in 2011 to create a demonstration, *Wastewater Garden*, that could grow both beauty and health in the southern Iraq marshes. I

Preliminary Wastewater Garden Designs, Southern Iraq Marshes, 2014

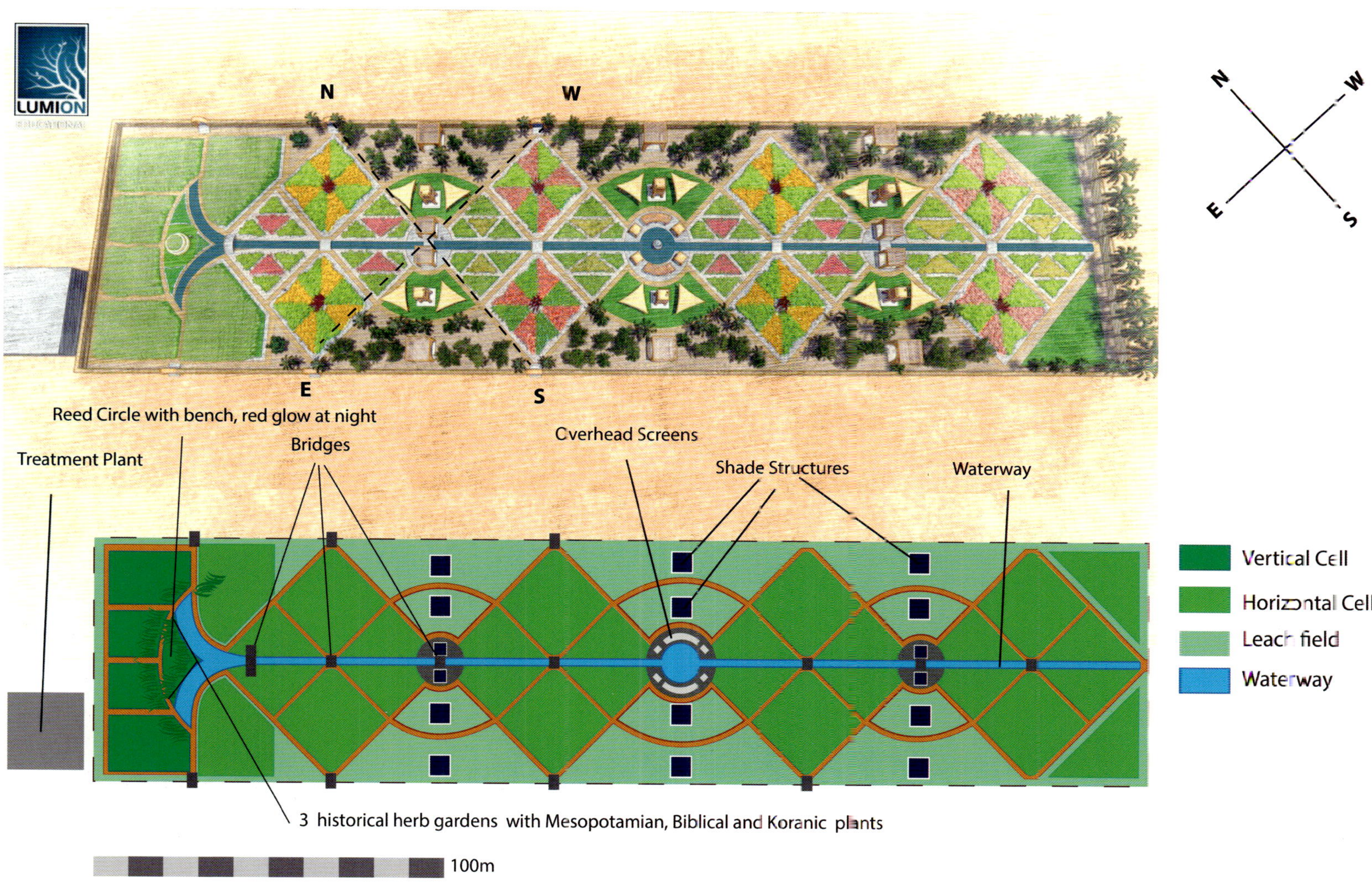

had recently learned from my New Mexico neighbors about growing gardens to purify human sewage. In the early 1990s they had created Biosphere 2, the first man-made closed biospheric system, as a laboratory in southern Arizona for studying the global ecology. Team member and environmental engineer Dr. Mark Nelson had developed the wastewater systems using natural plant and water cycles to recycle sewage. He has gone on to design and build hundreds of wastewater gardens for indigenous and other communities throughout the world. Mark agreed to go to Iraq with me to find the site of the Garden of Eden and re-create it as a symbol of our verdant future, turning waste into flora and providing crucial help for the Marsh Arabs of the historic Iraqi wetlands.

After our first trip to southern Iraq in 2011, Mark and I were joined in 2013 by an international team made up of an industrial designer, an archaeologist, and a second environmental engineer. We were funded through Nanyang Technological University, where I teach. By partnering with Nature Iraq, NGO, we have gained support at every level: towns, regions, and national ministries. We have been designing the wastewater garden and cultural heritage site for the largest marsh town, El Chibaish, on a twenty-thousand-square-meter site that will support approximately 7,500 Marsh Arabs. Our design uses local materials such as earthen brick, woven reed, and ceramic tile. Our sources of inspiration include Marsh Arab embroidered wedding-blanket patterns for garden design, Mesopotamian cylinder seal imagery for ceramic relief patterns, and marsh reed architecture for shade structures and viewing towers. In addition we will incorporate imagery of ancient flora and fauna from Mesopotamia as well as those

Flora Mentioned in the Koran, photographed at the Royal Botanical Garden, Amman Jordan, 2016

Mesopotamian Embroidered Woven Wedding Blanket, made by Women's Weaving Cooperative, Samawa, Iraq, artist's collection, 2014

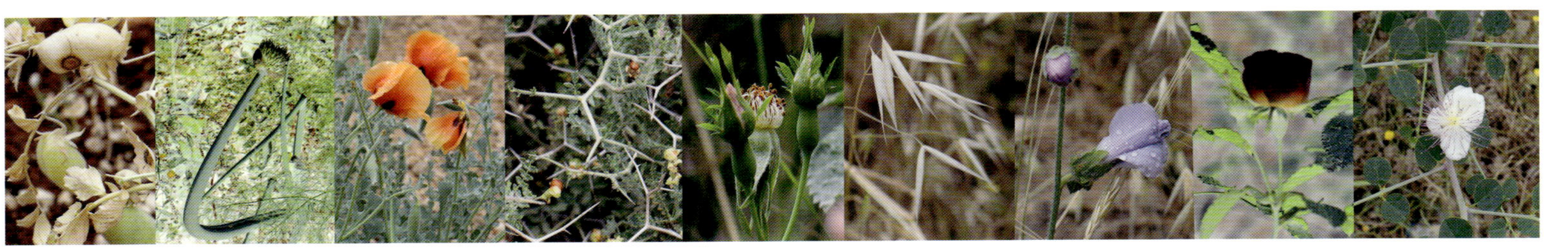

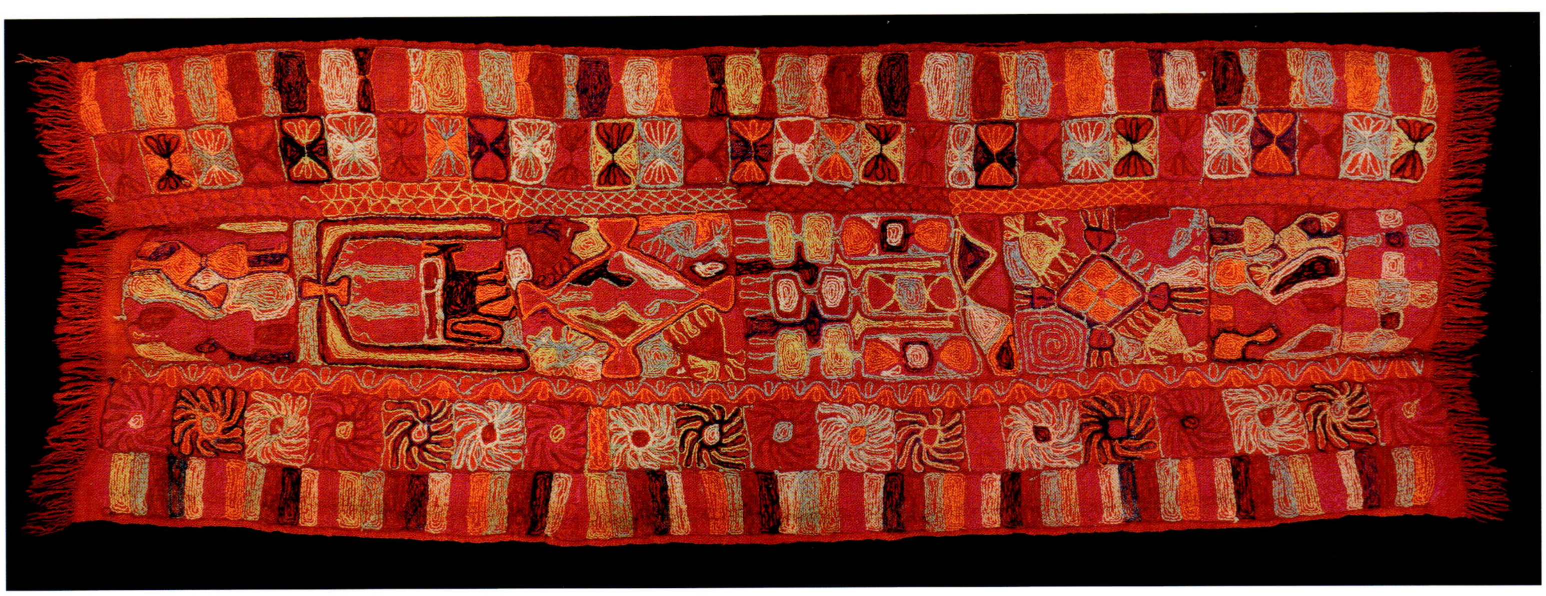

mentioned in the Bible and Koran. In the time after war, this will be a stopping-off point for tourists on their way to visit this newly designated UNESCO World Heritage Site and first Iraqi National Park, the Mesopotamian Marshland National Park.

In April 2016 I made a visit to the Royal Botanic Gardens of Jordan to photograph flora in the wild that were over five hundred years old and most likely would have been in Mesopotamia as well as in Koranic times—and perhaps in the actual Garden of Eden, if it ever existed. These flora are presented as a border in the photowork *Amethyst Room*. I hope someday to include plants from the manufactured Biblical gardens in Israel. The Jordanian Royal Botanic Garden wishes to keep species as pure as possible, so plants are protected in the wild and not cultivated. It is interesting how social terms that apply to migration and immigration apply to flora as well. When invading plants overtake, like Eucalyptus, they are unwanted. When they adapt and harmonize, like bougainvillea and olives, they can be accepted with grace.

Linen, tapestry, and metal are used as print media in this third section. The challenge for me as both an artist and an instigator of the *Eden in Iraq* project has been to create images that resonate, both conceptually and in a material sense, with the other two parts of *Eden Turned on Side*—*Photosynthesis* and *Volcano Cycle*—and that also remain companions to the *Wastewater Garden Project*. My intention for the images in this last section is that they be seen as weavings or tapestries, in the sense that different themes are being woven together: nature, culture, environmental and social history, and cultural heritage.

Beautiful rugs with abstract patterns from the natural world, prayer rugs, piles of blankets for overnight guests, and pillows on the floor are found in every marsh home. Thus our garden design is based on Mesopotamian wedding blanket patterns. In this landscape—exposed to extreme seasonal contrasts including flood, drought, violent rainstorms, sandstorms, scorching summers, and freezing winters—hanging textile panels convey a sense of wind and moving grasses.*

During my time in Iraq (from 2011 to 2015),

*My choice to use linen was influenced by the appearance of the earliest descriptions of making linen in hymns to Innana from Sumer sometime between 3500 and 1900 BC.

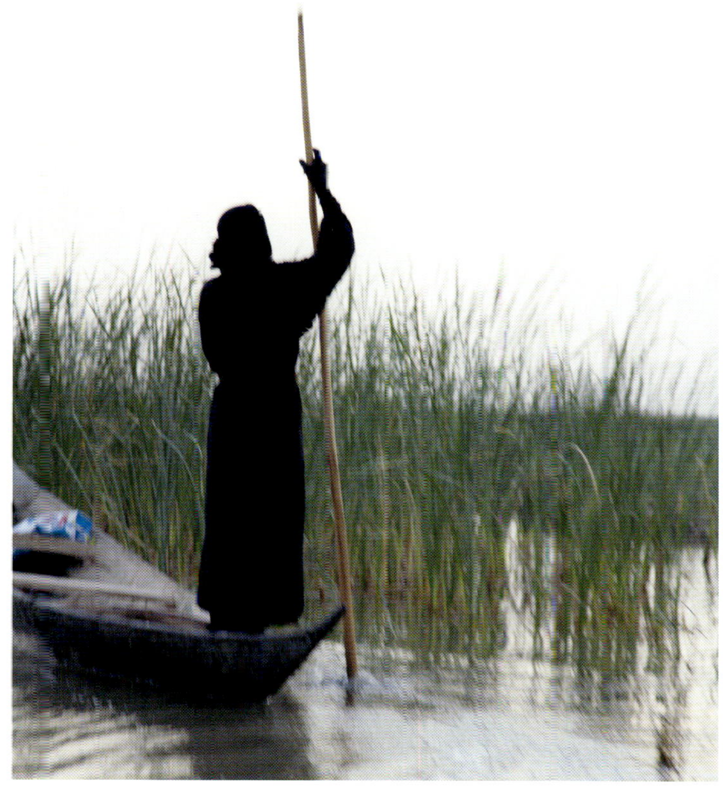

Marsh Woman Rowing, Southern Iraq Marshes, Iraq

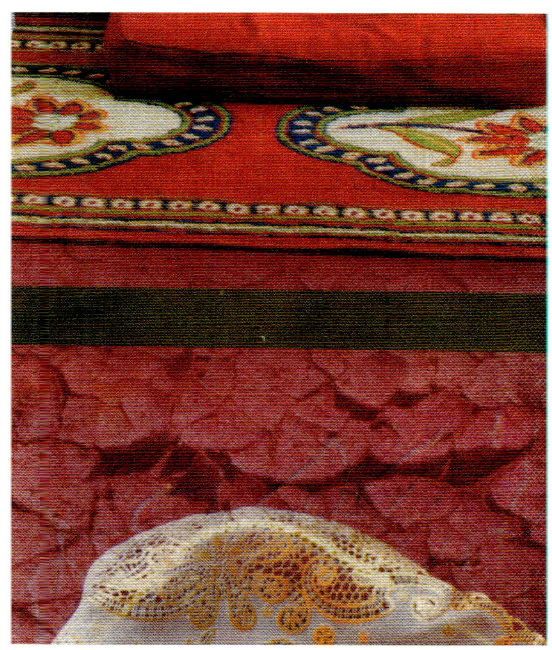

Linen detail

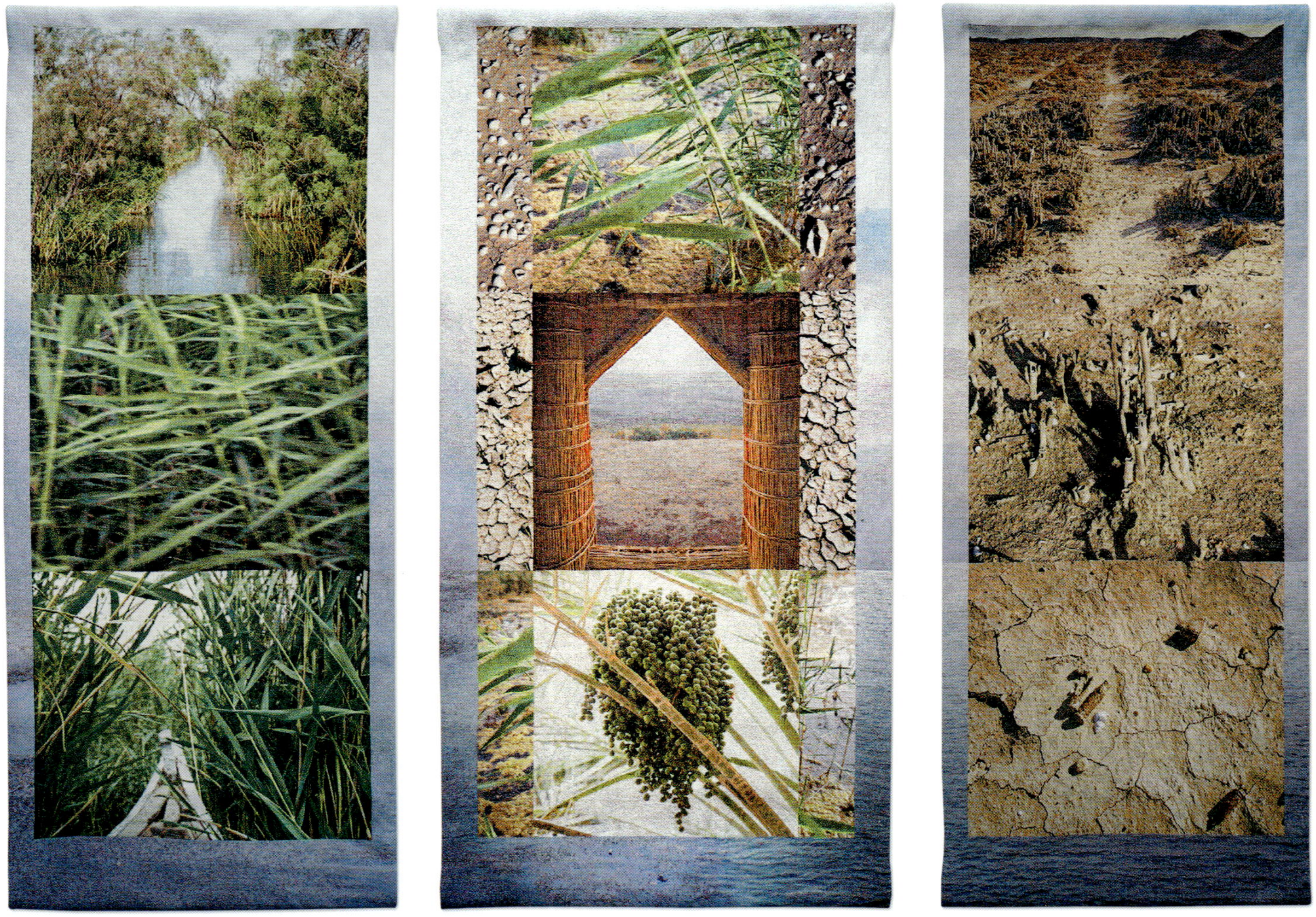

in the press of organizing meetings, being the only woman in a very male culture, driving long distances through parched desert away from the beautiful green of the marshes, I found myself searching for color. So when we would enter a home with beautiful amethyst curtains or a brilliant green kitchen, my eyes were refreshed by these oases. I seemed to be searching for the colors of Eden and found them within homes, in the bright dresses hidden by womens' abiyahs, and in the flora of the marshes.

Working in war-torn Iraq makes the simple idea of beauty very important. I could speak to town councils and families about beauty, about transforming the refuse from centuries of conflict back into a garden. I could offset the pain of finding bullet shells near a bombed-out shrine by juxtaposing those images with the beautiful, ancient snail shells that we found at the archeological sites of Ur and Uruk, far away from water.

Eden Turned on its Side looks at ecological and human processes across time that either reinforce or destroy the notion of Eden. This work focuses on poetic intersections of nature and culture in relation to ecological and social imbalance. Here people and nature are presented deeply connected, existing in threatened symbiosis. It is we who are on our sides needing to stand upright in the Garden.

Jacquard-woven tapestries hanging at Magnolia Editions, Oakland, California, 2017

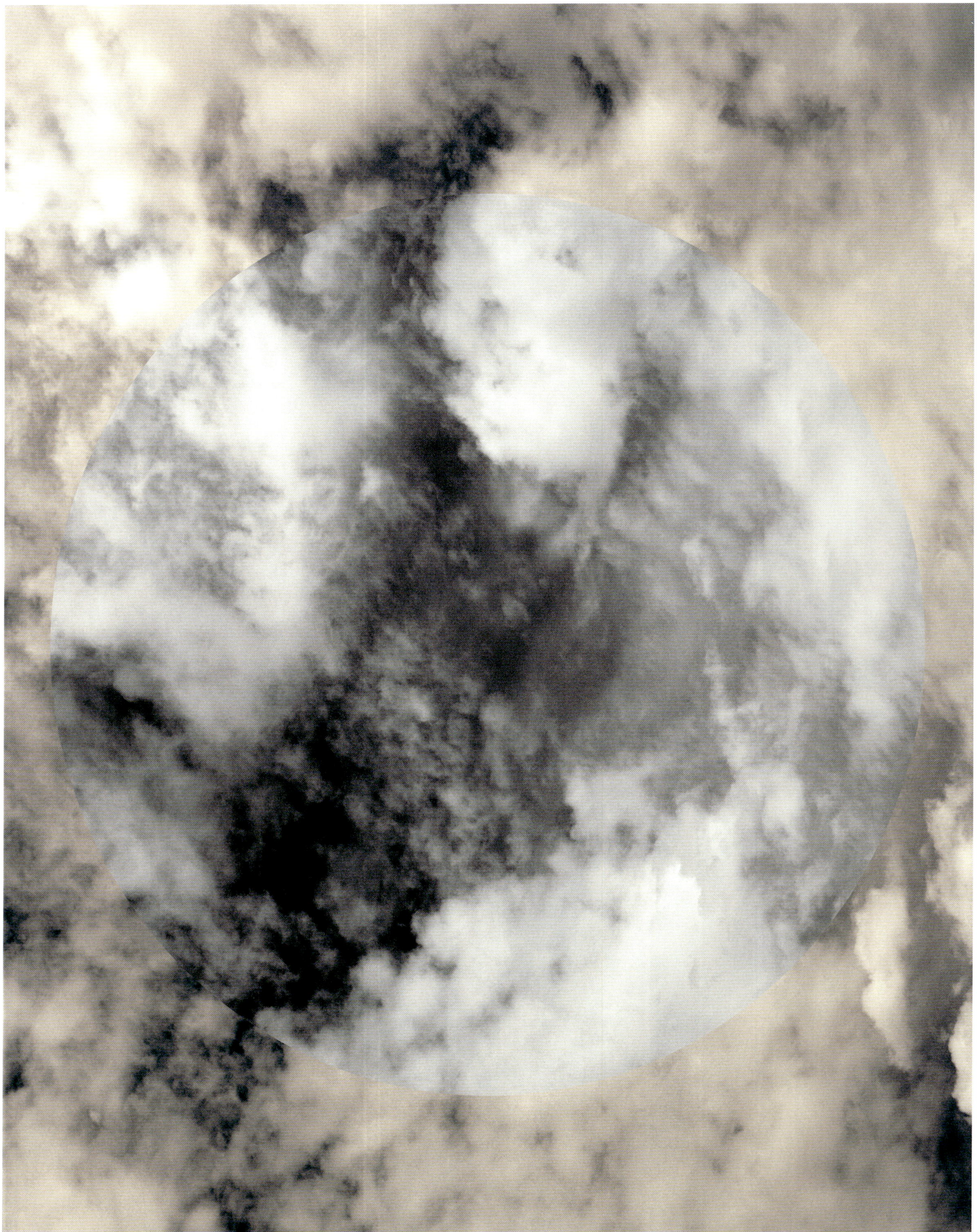

Acknowledgments

This book is dedicated to my dear friend, the late painter and environmental activist Ciel Bergman. Her example was a driving force behind this work. Ciel and her UC Santa Barbara art students spent a semester in 1987 cleaning the Santa Barbara bay beaches of refuse, to create an installation from the plastic laden garbage at the Santa Barbara Museum of Art Contemporary Arts Forum. Shortly after, Ciel had a dream that all that recycled plastic could be mixed with asphalt to pave roads. She gave the idea to her inventor-partner and engineer son, who spent fourteen years trying to make it a reality. Pavement tests with the New Mexico Highway Department still show Plasphalt™ to be a stronger material. But the New Mexico concrete industry, in step with the New Mexico Highway Department, blocked the results from being released, and finally their company had to close. Now asphalt mixed with plastic is paving roads throughout India. While she didn't reap any material benefits, Ciel demonstrated that an artist's unconscious can become conscious and move mountains. For this vision, and for her belief that beauty and art can heal the environment, I'm forever grateful.*

I am so fortunate that Dr. Shawn Michelle Smith approached me to bring all three parts of *Eden Turned on Its Side* together for the first time for this book and its accompanying exhibition. Her interest in my work has been inspiring, and working with her has been illuminating as well as a joy. I'm deeply honored that she has found a place for my artistic concerns within her engaged inquiry into the history of photography.

I'm grateful to Dean Kymberly Pinder of the School of Fine Arts, then acting director of the University of New Mexico Fine Art Museum, for bringing Shawn and I together and for shepherding this exhibition in its early stages. I'm deeply grateful to Arif Khan, director of the University of New Mexico Fine Arts Museum, for ingeniously making our project a reality. His support has been critical and his professional standards an inspiration. Brilliant designer and University of New Mexico Press Art Director Lisa Tremaine saw this book as a reality way before I did. I'm forever grateful for her creative energy to bring this book into being.

Volcano Cycle: Ring of Fire between Heaven and Earth (2010–2014) has been generously supported by Nanyang Technological University in Singapore with a Tier 0 research award through the College of Humanities, Arts, and Social Sciences. Professor Chris Newhall and Research Volcanologist Antonius Ratdomopurbo of Earth Observatory Singapore at Nanyang Technological University contributed invaluable support and counsel at Earth Observatory Singapore and in the field. Pra Purbo had directed the Mt. Merapi Volcano Observatory for twenty-five years in Jogyakarta without incident. Chris had predicted the 1991 Pinatubo eruption in the Philippines, thereby averting any casualties there.

Dr. Dan Schrag, at Harvard's Center for the Environment, deserves heartfelt credit for insisting in 2008 on my addressing carbon cycles as poetic material in my work. He gave me the challenge of trying to bring carbon so close that we'd feel it as not separate from our own skins. My friend Dr. Allison MacFarlane, geologist and past chair of the Nuclear Regulatory Commission, gave me Skype lessons on Carbon Cycles, feedback loops, and geologic time. Thank you, Dan and Allison, for helping me begin. Conversations with my favorite photo model, writer Anne Pedersen, also helped to deepen my understanding of environmental science.

In 2010, as I prepared to teach a new studio course called "Art and Ecology," I read Alan Weisman's *The World Without Us*. This became my premier text,

**The Ciel Bergman Plasphalt™ archive resides at the Center for Art and the Environment at the Nevada Museum of Art in Reno.*

Summer Cloud, 2009–2011, 43 × 32 ¾ in.

corroborating for myself and awakening my students to what we have wrought and the power of nature to override us. I am thrilled that Alan has contributed to this book, as he is on the top of my list of environmental thinkers. Thank you, Alan Weisman.

Deepest bows to my *Eden in Iraq* team—Assistant Professor Peer Sathikh and Drs. Mark Nelson, Sander van der Leeuw, Davide Tochetto, and Eng. Jassim Al Asadi—for believing in an artist's idea and using their skills to help realize it. We received a Tier 1 research award through the College of Humanities, Arts, and Social Sciences from the Singapore Ministry of Education with Professor Peer Sathikh as Principal Investigator. The photoworks for *Eden in Iraq* have been created with the help of this award.

The School of Art, Design, and Media at Nanyang Technological University in Singapore has supported this work for the last nine years in all ways: financially, intellectually, and spiritually. Deepest gratitude to Professor Vibeke Sorensen, Chair; Associate Chairs (Research) Professors Michael Walsh and Andrea Nanetti; Lew Huey Shan and Ivy Poh for budget management; and Muhammad Mustajab Bin Mohamad, Senior Executive of Publicity and Outreach, for everything in every way. In addition Dean Alan Chan of the College of Humanities, Art, and Social Science (HASS) has been at the forefront of support for this work. HASS Associate Dean (Research) K. K. Luke insisted that we apply for our research grant. Without him, *Eden in Iraq* could never have continued. Director Dr. Michael Khor (Professor Khor Khiam Aik) and Manager Dr. Ng Foong Loo, Yvonne, of the Research Support Office offered necessary support to guide us in administration of the research award.

Curator Patricia Levasseur De La Motte did the first in-depth writing about *EdenTurned on Its Side: Photosynthesis* for the exhibition at Chan Hampe Gallery in Singapore in May 2012. Over nine years, many assistants, technicians, and students have helped with the fabrication of this work. My main assistants—Molly Wagoner in the United States and Quek Jia Liang in Singapore—know how crucial they have been to the realization of this work. Students Abed Trianto and Clarence Lam accompanied me to east Java and the Mesopotamian Marshes respectively. Steve Zeifman at Rush Creek Editions in Santa Fe, Josef Blazer at Blazing Editions in Rhode Island, and Kelvin Mun at AVS Technologies in Singapore helped to print the work. Its been an honor at the end to work with Nicholas Price at Magnolia Editions in Oakland to bring the textile pieces into reality. Photoworks in tapestry and on linen are in the works.

Artist Dana Hart-Stone provided invaluable guidance in studying printing on textiles. Thanks also to James Wood, Brittny Dayes, Kim Richardson, Jason Rodriquez, Constantine Alatzas, Chen Zhi Lin, Wee Sher Ying, and Tse Man Nga.

To Nature Iraq, I offer my deepest gratitude forever for helping me to make these photographs: to CEO Azzam Alwash, Haider in Bagdhad, Araz in Suliemaniah, Ahmed and Made in El Chibaish, and especially to the Jassim Al Asadi family, without whom none of this would have happened.

Hatem Tafour, chief botanist of the Jordanian Royal Botanic Garden, provided staff support, vast expertise, and knowledge for my three days of photographing there.

The support and input of my dealer Brian Gross, of Brian Gross Fine Art in San Francisco, makes this work glow as it has for the past thirty years. He has been there every step of the way, refining refining, as has the sculptor and poet Rosalyn Driscoll, who has guided this artwork and the present writing with her clear, graceful wisdom. The root and home studio Jerry R. West provides for this artwork is critical.

Lastly, Ben Shedd has kept me pointing in the right direction to complete this work with love, energy, and any humor I can muster. While he arrived in the middle, he can see the whole.

Meridel Rubenstein

Contributors

Meridel Rubenstein began her professional career in the mid-1970s, evolving from a photographer of single photographic images to an artist of large-scale installations. She is recognized for juxtaposing highly charged materials and concepts to champion an awareness of how we are connected to place.

Rubenstein has been a visiting associate professor at the School of Art, Design, and Media at Nanyang Technological University in Singapore since 2007. She maintains her studio in Santa Fe, New Mexico, and has exhibited widely, including at the Louvre in Paris, the Museum of Contemporary Photography in Chicago, and the List Center for the Visual Arts at MIT.

She has received fellowships from the Guggenheim Foundation and Harvard University and awards from the National Endowment of the Arts, the Pollock Krasner Foundation, and the Rockefeller Foundation. She was educated at Sarah Lawrence College in New York and did special graduate studies at MIT with the eminent photographer Minor White. She received an MA and an MFA from the University of New Mexico in 1974 and 1977.

Rubenstein's twenty-year monograph, *Belonging: Los Alamos to Vietnam, Photoworks and Installations, 1980–2000*, was published by St. Ann's Press in Los Angeles in 2004 and includes texts by environmental writer Terry Tempest Williams and cultural and art critics Lucy Lippard, Elaine Scarry, and Rebecca Solnit. James Crump, curator, publisher, and filmmaker, wrote the biographical text.

A traveling exhibition to document Rubenstein's *Eden in Iraq* Wastewater Garden Project will premiere at the National Design Center in Singapore in October 2017. It will feature drawings, models, art and documentary photography, and video work.

Shawn Michelle Smith is a professor of visual and critical studies at the School of the Art Institute of Chicago. She has published six books on the history and theory of photography and race and gender in visual culture, including, most recently, *At the Edge of Sight: Photography and the Unseen* (Duke 2013), which won the 2014 Lawrence W. Levine Award for best book in American cultural history from the Organization of American Historians, and *Photography and the Optical Unconscious* (Duke 2017), coedited with Sharon Sliwinski.

Journalist and author **Alan Weisman** has worked in more than fifty countries and on all seven continents. His reports have appeared in the *New York Times Magazine*, the *Atlantic Monthly*, *Harper's*, the *Los Angeles Times Magazine*, *Vanity Fair*, *Mother Jones*, *Discover*, *Orion*, *VICE*, and on National Public Radio. His 2013 book, *Countdown: Our Last, Best Hope for a Future on Earth?*, now in thirteen foreign-language editions, won the *Los Angeles Times* Book Prize and the Paris Book Festival Prize for Nonfiction and was a finalist for the Books for a Better Life Award. His book *The World Without Us*, an international bestseller now in thirty-five languages, was named Best Nonfiction Book of 2007 by *Time Magazine* and *Entertainment Weekly* and was the winner of the National Library of China's Wenjin Book Prize. It was also a finalist for the National Book Critics Circle Award, the Rachel Carson Prize, and the Orion Book Prize. His previous books include *An Echo In My Blood*; *Gaviotas: A Village to Reinvent the World*; and *La Frontera: The United States Border With Mexico*. A senior documentary producer for Homelands Productions, he lives in western Massachusetts with his wife, sculptor Beckie Kravetz.

Printed in Singapore
22 21 20 19 18 17 1 2 3 4 5 6

Library of Congress Cataloging-in-Publication Data

Names: Rubenstein, Meridel, photographer. | Smith, Shawn Michelle, 1965– writer of added commentary. | University of New Mexico. Art Museum, issuing body

Title: Eden Turned on Its Side / photoworks by Meridel Rubenstein ; foreword by Arif Khan ; essays by Shawn Michelle Smith and Alan Weisman.

Description: Albuquerque: University of New Mexico Art Museum, 2017. | Includes bibliographical references.

Identifiers: LCCN 2017017338 | ISBN 9780826359179 (cloth: alk. paper)

Subjects: LCSH: Nature photography. | Photography, Artistic. | Environmental degradation—Pictorial works. | Nature—Effect of human beings on—Pictorial works. | Human ecology. | Eden in art.

Classification: LCC TR721 .R79 2017 | DDC 779/.3—dc23 LC record available at https://lccn.loc.gov/2017017338

Cover photograph
Volcanic Leaf, Mt. Merapi eruption, East Java, Indonesia, 2011

Section divider photographs
Photosynthesis Leaves, 2012
Lava From Mt. Merapi, 2010 and 2012
Ancient Shells and Ceramic Peg from Ur and Uruk (Over 4,000 Years Old)

Designed by Lisa C. Tremaine

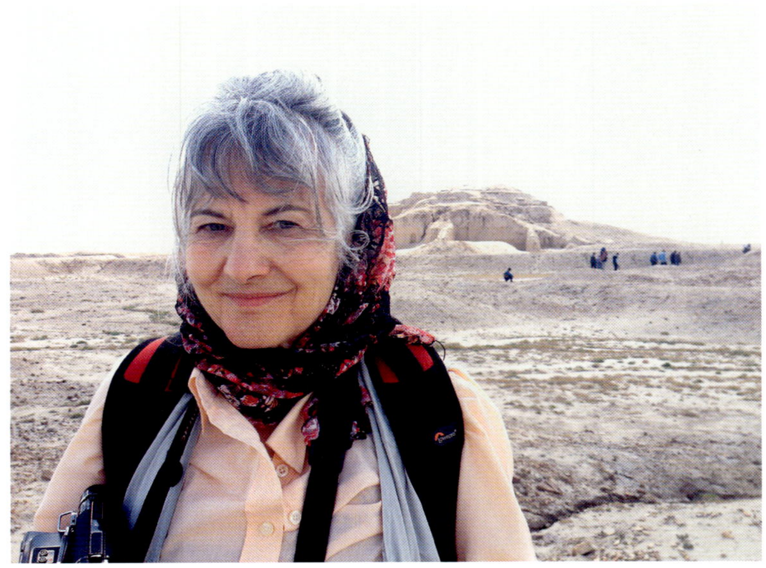

Meridel Rubenstein at Temple of Inanna, Uruk, 2014, by Lori Schwilling

Meridel Rubenstein photographs in a variety of ways, mostly capturing images on large- and medium-format Ektacolor film, as well as using digital files. The film is scanned on a high-resolution film scanner and then either printed onto the substrate or digitally modified and/or montaged, and then printed onto 100 percent rag papers, aluminum, or linen fabric prepared specifically for inkjet pigment inks as well as jacquard tapestries with the image woven directly into the material.